# FRAMING FLORAL TECHNIQUES

# FRAMING FLORAL TECHNIQUES

## Floral Design Skill Building, Inspirations & Explorations

### Renee Tucci AIFD PFCI

SCHIFFER PUBLISHING

4880 Lower Valley Road • Atglen, PA 19310

# Acknowledgments

I get by with a little help from my friends.

For their technical review, valuable advice, unending support, and simply helping me to realize that this book was possible, I would like to acknowledge:

Tim Farrell, AIFD, AAF, PFCI

Darcie Garcia, AIFD

Marisa Guerrero, AIFD

Dana Marston

Matthew Ross

Tara Stemborski

Shannon Toal, AIFD

Michael Tucci Jr.

Accent Décor

For Michael

Who makes my crazy ideas seem sane, and never lets
me forget the importance of a good book and a wallet
full of library cards.

# Contents

60

64

68

74

82

86

94

90

100

104

## Connections . . . . . . . . . . . . . . . . . . 58



## Organized . . . . . . . . . . . . . . . . . . . . 80

## Revealed . . . . . . . . . . . . . . . . . . . . 108



## Covered . . . . . . . . . . . . . . . . . . . . . . 146

174

178

190

184

204

194

200

210

216

# Take Imperfect Action!

As a lifelong learner, consumed with insatiable curiosity, the hows, the whats, and the whys dominate my thoughts when discovering a new-to-me floral design technique or style. These questions spur me into immediate action: first thinking about the process, and then physically taking this new floral experiment for a spin. Most often, these first-try projects do not end in success, and they are generally followed by more questions, a reexamination of the steps and possible pitfalls, and sometimes, thoughts of serious self-doubt. But then I remind myself, things are always messy in the middle. The key is to keep pursuing and to reach the other side.

That pursuit of taking imperfect action is how we build our metaphorical toolbox. It's full of horticulture habits that we've all learned from a young age. It also holds the techniques and ways of thinking that let you bring new innovations to life. We know that roses have immense beauty, but they also have thorns. Can those thorns be used as a piercing instrument in an interpretive creation? Freesia has a sweet and peppery fragrance that many have harnessed for perfumes and body lotions. Can that fragrance be layered with other aromatic flowers to create a new, never-before-sniffed scent? Rainbows enchant children of all ages. How can the colors of the rainbow be mixed to build expressive color palettes? The answers to these questions lie within the tools in your toolbox.

My directive for you, my flower-loving friend, is to take imperfect action: to enter the arena, to try new things. Find a technique that you have never explored, or previously pursued and abandoned, frame out your approach, and dive right in. Push the limits of your comfort and rise above the idleness. Discover easier and more practical ways to accomplish techniques, and master those that you have already been practicing.

Then, share what you have learned with the world. Fight the urge to wait until everything is perfect to take the next step. Perfect will come when you have put your own spin on what you have learned from someone else. Perfect will come when you have inspired someone else to take imperfect action. Perfect will come when you make the leap from one idea to another because you have simply inspired yourself.

Explore, innovate, and take imperfect action!

## Framing Floral Techniques on YouTube

Throughout the step-by-step processes in the pages that follow, you will see this QR code featured. Scanning this code will send you directly to a special Framing Floral Techniques playlist on YouTube. There, you will find short videos that show some of the beginner building blocks needed for designing, expansions of some techniques, and alternative outcomes.

## Substitutions Are Smart!

Don't be afraid to sharpen your skills with substitutions. If the materials mentioned in the project lists aren't readily available, look for materials that mimic the form, color, and style of the missing pieces. A carnation is a great substitute for a rose, an alstroemeria is a great substitution for a tulip. Don't let missing materials hold you back from practicing projects!

## Basic Tools and Mechanics

1. Knife
2. Ohana clippers
3. Bunch cutters
4. Wire cutters
5. Needlenose pliers
6. Floral glue
7. Pan melt glue
8. Spool wire
9. 18" stub wire: 18, 20, and 24 gauge
10. UGlu dashes and strips
11. Foam: traditional green brick, Midnight, Sahara
12. Stem wrap tape
13. Waterproof tape

# Foundations

## Basing

All good things rise from a strong foundation. When that foundation is a thing of beauty, well, that's just a bonus! The list of treatments that can be created at the base of a design is as long as your imagination is big. When flowers emerge from a divinely decorative, totally textured, lusciously layered base, the result is utterly unique.

## Clustering

When many individual pieces of the same material mesh together, it creates melodious music. The distinct, individual pieces disappear into the mass of the message, leaving only the effect to enjoy.

## Pillowing

Like making waves with your hand out of the car window into the rushing wind, pillowing takes the eye up and over each magical mound of flowers, allowing for slight rest in between the mini mountains.

## Terracing

To build an in-depth base for your design, try terracing. Like a sensational staircase, components are cast one piece just in front of the next. This pleasing pattern pulls the viewer in to the feast of the focal area.

# Pearly Petals

Providing an interesting base for your design can lead to serious success when using minimal materials. Foliage, petals, gems, and moss are the most common basing materials, but here we have stepped outside the "common" box. Creating an extended column above the cylinder vase allows ample room for the layering of pearls onto the surface of the carved-out floral foam, where iridescent pillows give way to humble pink roses and delicate maidenhair fern.

## MATERIALS

- Bear grass
- Equisetum
- Hot-pink roses
- Maidenhair fern
- ½" diameter pearls
- Floral foam
- Hot glue
- 6" clear cylinder vase
- 20-gauge wire

### Tech Talk

* The pearls on the base have been *layered*, leaving no space between them.

* Due to the naturally round shape of the pearls, a *pillowing* effect is created on the base.

* The stems of equisetum have been *tailored* to specific heights, and even cut at specific angles, to create an interesting column structure.

**1.** Thread equisetum stems onto 2 pieces of 20-gauge wire to form a mat. For a 6" × 6" cylinder vase, the mat should be 16" long. The equisetum can be left at its natural length, to be trimmed and tailored later.

**2.** Curve the mat into a cylinder and place it in the vase.

**3.** Working with a presoaked brick of floral foam, set it next to the vase in the upright position for an easy measure of where to trim the equisetum. This will be the first, rough cut of the equisetum to make it easier to add the foam in. The final, tailored cut will be made in the following step. Trim the equisetum, leaving it extend about an inch above where the top of the foam will be. Then, cut a "U" shape out of the front portion of the equisetum to allow for more of the surface of the foam to be accessed. Shave the corners of the floral foam as seen in the Stacked Steps project (page 26), and place the upright brick of wet foam in the vase.

### Tech Tips

* Using wire cutters, cut the ends of each piece of wire on an angle for easier piercing and threading of the equisetum.

* Try to choose equisetum stems that are as straight as possible.

* The sloped equisetum stems around the front benefit from not just a straight cut, but angling the cut downward, exposing a teardrop shape from their hollow stems.

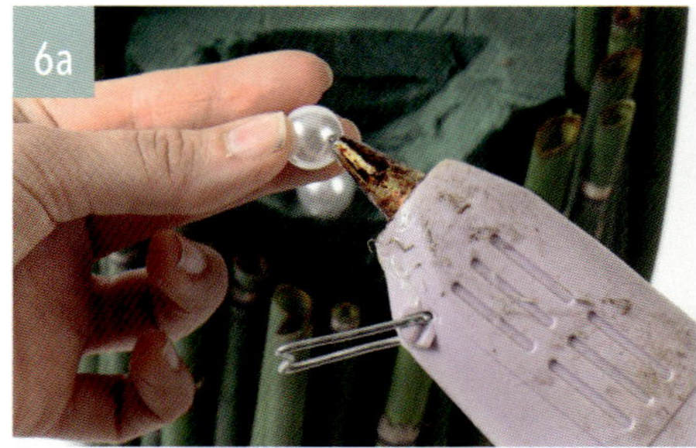

**4.** Finish trimming the equisetum so that it stands about 1½" above the foam in the back and slopes down to reveal about 2" of exposed foam at the front.

**5.** Using a knife, carve out a well or depression from the front half of the foam.

**6.** Apply a dot of hot glue to a pearl and fasten it to the carved well area. Repeat to cover the front of the foam with a layer of pearls, then continue to mass the pearls higher to fill the well while keeping the angled, or spilling, effect.

**7.** Place the roses into the back half of the foam. Arrange the roses using sequencing: place the tightest blooms highest, moving down to finally place the most widely open rose.

**8.** Add maidenhair fern around the base and through the cracks of the equisetum into the foam, tumbling down the column to introduce a fluffy texture in contrast to the very repetitious and smooth equisetum.

**9.** A grouping of bear grass is added to the back to bring the design into proportion and to balance the tumbling maidenhair fern.

# Clustered Calm

Clean and distinct design can bring calmness to a visually fatiguing world. By sectioning stems of foliage and adding them into this wreath so close together that you cannot quite make out the individual placements, a degree of visual peace can be accomplished.

## MATERIALS

◇ Aspidistra leaves

◇ Green cymbidium orchids

◇ Green dendrobium orchids

◇ Green trick dianthus

◇ Leatherleaf fern

◇ Ming fern

◇ Sprengeri

◇ Tree fern

◇ Variegated pittosporum

◇ 22" mâché-backed floral foam wreath

◇ Wire easel

◇ Picked water tubes

◇ UGlu dashes

***Tech Talk***

✳ The *color* palette in this piece is *monochromatic*.

✳ The clean and *repetitive* effect of the aspidistra leaves, emerging from their *pattern* of placement around the perimeter, creates a line that the eye will follow as it travels around the wreath.

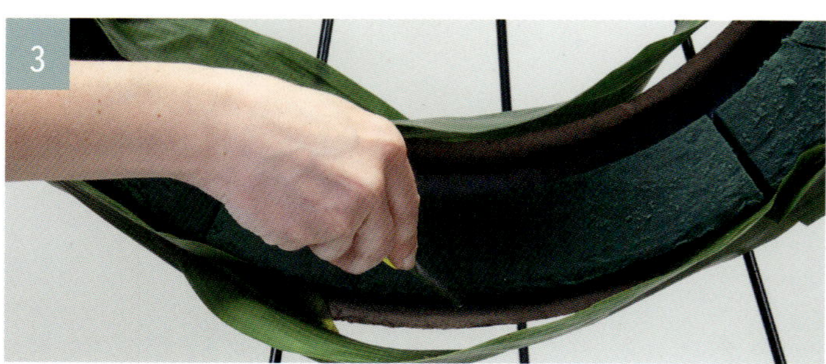

**1.** For ease of design, set your wreath on a wire easel that stands at a comfortable height for you. Shave the bottom half of the central vein from each aspidistra leaf as shown in the Sheltered Sanctuary project (page 156).

**2.** Place the shaved aspidistra leaves around the outer and inner circumference of the form, overlapping their ends as you insert each new stem through the previous leaf's tip to pin it down.

The final leaf tip can be tacked onto the first stem with a UGlu dash.

**3.** Using a knife, score the foam to create sections where you will add each flower cluster. A template can be helpful as you plan the placements.

**4.** Start by placing a few stems of each variety in their sections, so that you can envision the color and texture balance. Insert your variegated pittosporum in two opposing sections to lighten the overall color balance and, because it is the most starkly different foliage, having two sections of this will help establish visual repetition. Adjust as needed until you are happy with the pattern and then proceed to fill the wreath.

**5.** Add sprengeri and ming fern. Be aware of the thorns on both varieties; they will leave you with painful souvenirs embedded into your fingertips for weeks.

Add the delicate tree fern by cutting individual laterals into short, small, and bundled clusters, and placing them into the foam at the same time.

Green trick dianthus has a natural pillow shape. Maneuver each flower head to blend and mesh into the flower next to it to achieve the clustered look.

**6.** The laterals along the main stem of the fern can be cut into smaller, individual pieces, the bottom of their stems stripped clean, and added in. These leaves come to a point, so choose their alignment direction and follow it with placements to help create a visual path for the eye to follow.

**7.** Where the aspidistra leaves naturally want to bow away from the wreath form, allow your materials to fill in by leaving the stems longer to fill the gaps.

**8.** Once the clusters of greens are all added, create a focal point with 3 green cymbidium orchids that have been placed into water tubes.

**9.** Last, create a secondary, more subordinate emphasis area by trimming the open bloom of the dendrobium orchid stems into smaller sections and adding them in tightly, creating a grouping.

### Tech Tip

* To create the predetermined section for each material, you can also use the breaks in the foam that are in the manufactured form.

* Cut a slit in the top of the water tube by pushing your knife into the access hole. This allows the thicker orchid stem to slide in easier.

# Bunch Bouquet

A unique and modern bouquet can be easily designed by using the clustering technique with stems of flowers. In this bouquet, smaller bunches have been created by gathering like stems together so closely that, with the exception of the ranunculus and their distinctive graduating centers, it is difficult to decide where one stem stops and the next begins. Once these miniature bouquets are bound, they can be assembled together and wrapped as one larger bunch bouquet.

## MATERIALS

- Lacey dusty miller
- Lavender lisianthus
- Pink carnation
- Pink peony
- Pink ranunculus
- Purple hydrangea
- Purple trachelium
- Cable ties
- Coordinating ribbon
- Bind wire
- Floral pin

### Tech Talk

* The more mounded you make the clustered bunches, the more you will achieve a *pillowing* look between the bunches. Whether the aim is to achieve this look or not is up to the designer!

* Adding an element like dusty miller between each bunch will bring *unity* to the overall bouquet.

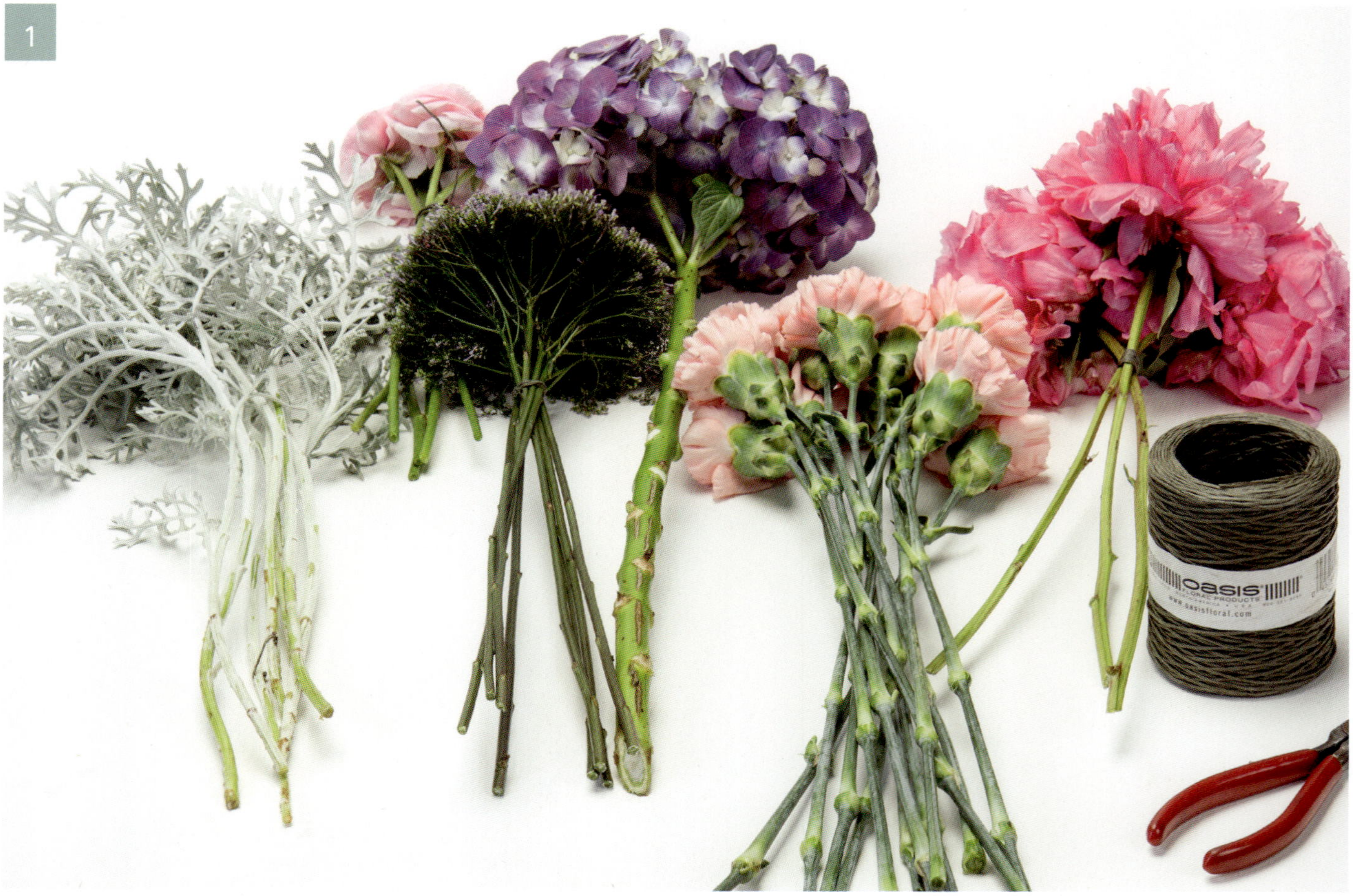
1
OASIS
FLORAL PRODUCTS

2

3

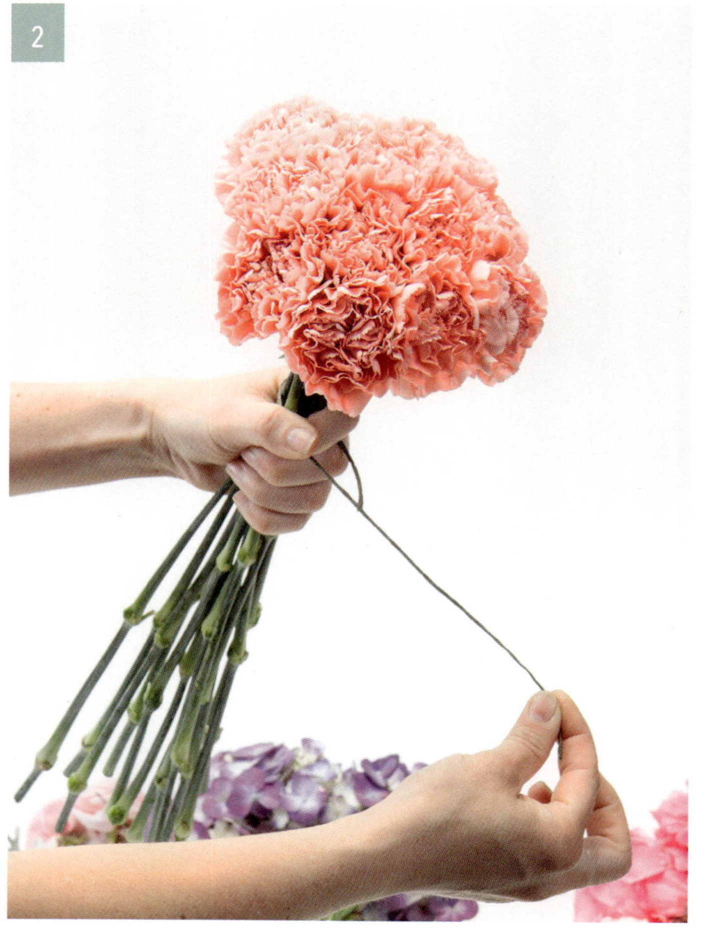

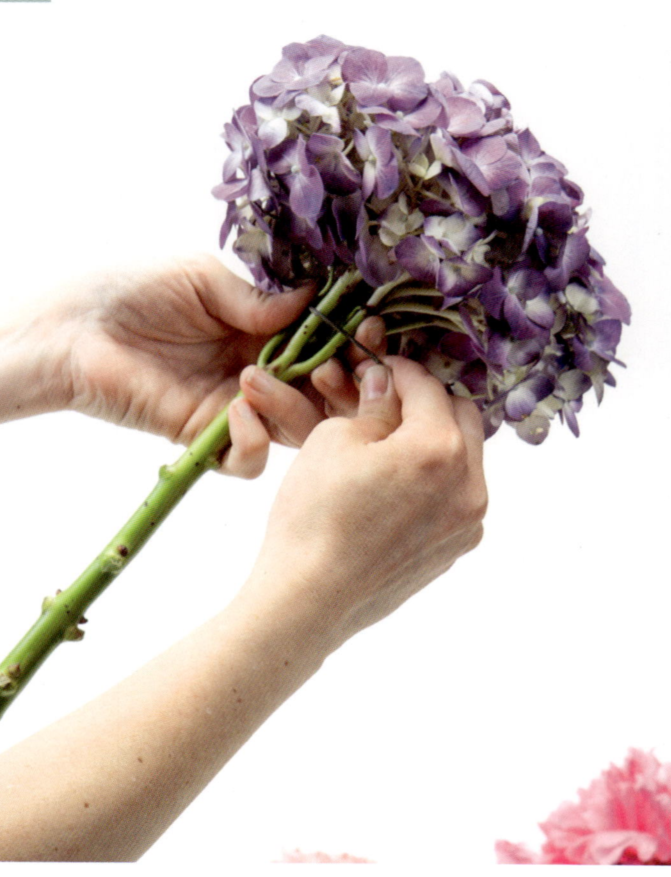

**1.** Gather each type of flower together and prepare the stems by stripping any foliage, lower laterals, or thorns.

**2.** Shape each cluster of stems so that they create a full, mounded, and compressed look, then bind together with bind wire.

**3.** For the hydrangea, gently bind the blossoms even tighter with a length of bind wire.

**4.** Start to place the bunches together, playing with their positions and placements, and working out how best to fit all of the puzzle pieces together to create a pleasing combination of forms, colors, and textures.

**5.** Fan out and place a stem of the dusty miller between each cluster to bring in color contrast, and to enhance and emphasize each individual bunch.

**6.** For a temporary hold, bind the stems lightly with a cable tie.

**7.** Cut a length of ribbon about 2 yards long. Starting with the middle of that length of ribbon just under the flowers, work your way down, creating a crisscross/twisting effect, or what is commonly known as a ballet/ballerina ribbon wrap. End three-quarters of the way down the handle with a square knot. Secure the knot in place with a floral pin, pushed in on an upward angle so that the tip of the pin gets lost in the center of the stems, rather than straight through and out the other side.

**8.** Snip and remove temporary cable tie.

**9.** To finish the tails, fold them in half, widthwise, and cut the ribbon on an angle, or a "V" shape. When unfolded, you will see a fishtail finish.

**10.** Trim stems flush and neatly, 2" longer than the ribbon wrap.

# Purple Pillows

Because sometimes you just want to pull up a pillow and rest your head on those beautiful botanical blooms! The pillowing technique is a lot like it sounds: creating domes or pillow-shaped mounds by clustering flowers together. This is a great way to create a clean, sophisticated, but substantial look at the base of your design, and it can also be done in taller form for larger compositions.

## MATERIALS

- Inchplant
- Lavender cremone chrysanthemums
- Lavender lisianthus
- Lavender roses
- Nandina
- Nerine lilies
- Purple hydrangea
- Floral foam
- Poly foil
- Green split peas (dried)
- Triple brick mold
- Wired wood pick

### *Tech Talk*

* By carefully assessing each blossom, you will see that even something like a rose, which blooms on all sides, has a face and can be placed in a way to maximize the viewers' enjoyment, and to help you create the *form* you are going for. In this case, it is helpful to use the *facing* technique for both the cremones and the roses to get a distinct pillow shape.

* The addition of the nandina in this composition not only increases the physical *size* of the design, but it employs greater positive, and negative, *space.* Draw an imaginary box around the entire composition and you will see all of the empty areas that have become a part of the design. These areas void of material are of equal importance.

**1.** Line the brick molds with poly foil, creating a temporary liner to keep moisture in. Once wet floral foam is added, trim excess foil, leaving about a ½" lead. Then fold that down and around the brick.

**2.** Trim the lateral florets from hydrangea head, leaving the stems as long as possible. Put them back together into the foam, with even more precision than Mother Nature did when she grew the hydrangea bloom. Since hydrangea require plenty of water, be sure to anchor them as deeply as possible into the floral foam.

**3.** Add the lavender cremone mums, paying careful attention to gentle bends in the stems and the direction that the flower head is facing. Use that face/direction to your advantage, to help strengthen the dome shape you are creating. Be sure there is a distinct end to where the cremones end and the hydrangea begin, creating the pillowing effect.

**4.** When adding the nerine lilies, you will use the same technique as with the hydrangea; cut the blossoms from the main stem. Then place them back together as one even more perfect pillow than they naturally grew in.

**5.** Repeat the pillowing with all materials until you have filled the molds. Stagger the placement of your pillows; building some toward the front edge, some toward the back.

**6.** Your nandina branches will most likely need to be edited so that just the right amount of branching stands above your blooms. Look at each branch with a keen eye, imagining what will be left if you cut this lateral or that, and trim as needed.

**7.** Cut tendrils off of a potted inchplant. For more leverage when adding to the foam, attach a wired wood pick to the base of each stem, as shown in the Stacked Steps project (page 26). Drape the tendrils over the mold edges.

**8.** Top any exposed floral foam with dried green split peas. Press the peas into the foam gently to get them to take hold and to keep from shifting.

### Tech Tip

When you deconstruct only to reconstruct, as we did here with both the hydrangea and the nerine lilies, you get the most control over creating the mounded shape you need. Pick and choose which laterals serve the composition's needs, and arrange the angles, meshing or tangling the florets into each other as needed.

# Stacked Steps

Just like rocks stair-stepping down a cliff, your flowers can create depth in a grand and gradual way. Here we have focused on freesia as our primary blossom, which, as Mother Nature intended it, blooms in a terraced fashion. By bringing each bloom just slightly in front of the previous, we are able to boost the terracing effect exponentially.

## MATERIALS

- Aspidistra leaves
- Bear grass
- Lavender freesia
- Lavender airbrushed plumosa
- Lily grass
- Variegated lily grass
- 20-gauge stub wire cut into pieces, or greening pins
- Midnight floral foam
- Wired wood pick
- Footed pedestal

### Tech Talk

* True *rhythm* can be developed with proper placement of the grass bundles. Allow them to dance around each other, to intertwine, and to weave over and under each other.

* While there are many *dynamic lines* created by the grass in this piece, the strongest line is *static* and is created by the freesia. It helps calm and center the eye and leads it down the intended visual path.

**1.** Shave the corners of a brick of wet floral foam and set on the pedestal.

**2.** Prepare the aspidistra leaves by cutting the stem off and shaving the bottom half of the central vein from each, as shown in the Sheltered Sanctuary project (page 156). Lay an aspidistra leaf over the top, leaving the ends hang over. Prepare a few pieces of floral wire bent into a "U" shape, or grab a handful of greening pins.

**3.** Using the pins, tack each side of the leaf down and to the foam. Then, similar to a gift-wrapping technique, fold and tuck the corners flat against the foam and pin them into place as needed.

### Tech Tips

* Midnight foam is best used in this project and any other where the foam may show at the completion of the design. This foam is gray when dry, turns black when wet, and recedes into a composition, making it harder to see if any foam mechanics are not covered. Additionally, this foam can be purposely left exposed and become a decorative part of the design for a contemporary arrangement.

* There is a front and a back to a blade of grass. When lining up your bundles, be sure the blades are all lying in the same direction.

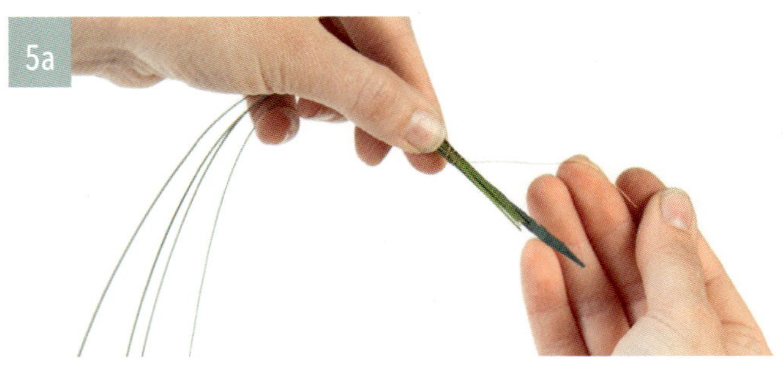

**4.** Lay a second leaf around the top/side of the foam, aligning the leaf edges a bit diagonally and wrapping downward, pinning as you go. Continue with wrapping the foam until the brick is predominantly covered in aspidistra leaves.

**5.** Begin wiring your grass bundles by aligning the tips of about 4 pieces of grass. Lay the tips of the grass parallel to the square end of the wood pick and bind them together with the attached piece of wire. Follow the blades down until you feel the first blade get thicker at its base: this is where all 4 blades should be cut. Then, bind the base of the blades to a wired pick so that there is now a pick on each end of the bundles. Prepare several bundles of lily grass, variegated lily grass, and bear grass in this way.

**6.** Begin adding your grass bundles, one at a time, by piercing them through the aspidistra with the pointed side of the wood pick until the pick is buried within the foam. Then wrap the grass around the tower and push the second pick into the foam when you run out of length. Continue adding your bundles this way, creating a random pattern, evenly spacing your variegated bundles to create interest.

**7.** If you find it a struggle to pierce the pick through the aspidistra, cut a small slit with your knife first, and push the pick in through this entry point.

**8.** Prepare several bundles of bear grass that have been braided to create even more texture and visual interest.

**9.** Next, add freesia stems, beginning on the top/ left area of the brick and working forward on an angle. As you progress down the brick, place each bloom so that it is slightly longer than the previous, creating the terracing effect.

**10.** Use the freesia's natural direction to your advantage as you arrange them. In order to help blend the flowers together, you may need to edit some of the larger/ bottom blooms from the stems as you place each piece.

**11.** You need to anticipate your spacing needs as you work down the design. To create a cascading energy, the top of the freesia line should be compact, and the freesia clusters should be placed more loosely and spaciously as they descend.

**12.** Finish the composition by adding contrasting texture with plumosa that has been airbrushed the same lavender shade as the freesia.

# Adorned

## Airbrushing

To paint, to tint, to shift color, to shade, to airbrush . . . they all direct the designer to pick up their favorite can of colored floral spray paint. To enhance the natural nuances that the flowers are born with, or to slather a serious statement for a bold composition, simply add your chosen hue straight from the can!

## Facing

Let's face it, we all want to put our best foot, or in this case our flower's face, forward. It's clear which way a sunflower, an anemone, or a gerbera daisy want to face, but it requires closer study to find the face of a rose or a hydrangea. By twisting and turning the stems, by studying the folds in the petals and the curves of the blossoms, we can make sure that basking in their beauty is easy for the beholder.

## Layering

Sometimes laying it on thick can be not only a good thing, but also a way to bring a composition from great to grand. Overlapping materials and creating a three-dimensional appearance by means of perfect placement or supportive mechanics allows for design depth and intricacy.

## Leafwork

When precise and repetitive placement of a foliage covers a container, you won't "be-leaf" the effect that can emerge! Scales, patterns, and surreal textures materialize when leaves are reimagined onto the form of a creation's container.

# Sunflower Sensation

There are a few flowers that shine best when they dominate a design, and sunflowers fit into that category. Their striking color and form are universally known to be a symbol of happiness, so why not multiply that and create a sensational project filled with them, but with a twist! By boosting the sunny petals with an airbrushing technique, we can create a graffiti, or variegated, look that will cause the viewer to do a double-take.

## MATERIALS

- Aralia leaves
- Aspidistra leaves
- Birch branches
- Bleached ruscus
- Buplurem
- Kiwi vine
- Medium and large sunflowers
- Nigella pods
- Pink astillbe
- Pink matsumoto asters
- Pussy willow branches
- Queen Anne's lace
- 9" design dish
- Grande floral foam brick
- Chicken wire
- Chenille stem
- Tall decorative vase
- Yellow spray paint
- White floral spray paint

### Tech Talk

* It's clear that this is a SUNsational design! The sunflowers are the *dominant* material throughout. The accent fillers and pink asters help to balance out the heavier sunflowers and bring in layers of *texture*.

* Keeping your elements *grouped* makes for a cleaner, more organized design. It also helps the viewer really appreciate each element as their eyes travel through the composition.

* When creating an asymmetrical design, such as this, physical and visual *balance* is especially important and can be achieved with botanical materials and/or the employment of *negative space*.

**1.** Lightly airbrush both sides of the bleached ruscus with the yellow paint.

**2.** Create your base by setting the fitted, wet Grande brick of floral foam onto the design dish and wrapping it with chicken wire for added stability.

**3.** Place several birch branches in the foam on the left, grouped together. As these are added, be sure to look at the laterals on the branches, which direction they are pointing, and edit as needed to keep a strong and clean line.

**4.** Balance the birch with pussy willow branches on the right side, slightly shorter.

**5.** Insert kiwi vine, grouping them, and angling them downward on the lower right. Choose stems that fit together to form a spilling look from the point of origin. Read more about point of origin in Summer Classics project (page 44).

 Check out this video for more how-to steps for preparing the base dish for this project.

**6.** To airbrush your sunflower petals, bend a chenille stem to increase its surface area. Spray white paint onto the chenille. Quickly, while the chenille is still wet with paint, brush the paint onto the sunflower petals in a random pattern.

**7.** Create a focal area in the lower/central area of design with the large airbrushed sunflowers.

**8.** Mingle the painted ruscus with the birch branches in the upper left to bring the color up into the striking line, and to balance the visual weight of the kiwi vine on the lower right.

**9.** Add medium graffiti-airbrushed sunflowers around the focal area and spilling out of the lower right, among the kiwi vine. Alternate their depth for visual interest.

**10.** Frame the design with aralia leaves.

**11.** To quickly cover the open area of the oasis, trim the bottom of an aspidistra leaf on a sharp angle. Then, fold the angled cut stem back into the center of the leaf, piercing through and creating a natural, mechanical hold: an aspidistra ribbon. Add several of these into the center foam area.

To see this aspidistra trick in action, scan this code for a video.

**12.** Enhance the aspidistra area with accent fillers pink astillbe and Queen Anne's lace on the left, keeping them grouped for maximum impact.

**13.** Balance the airy accent fillers with buplurem on the right.

**14.** For color balance, add a grouping of pink asters among the buplurem on the right.

**15.** For more visual balance, add nigella pods on the left and edit ruscus stems by clipping out heavier laterals.

**16.** Finish the back with a clustering of short greenery to complement and cover mechanics. Israeli ruscus, pittosporum, or more rolled aspidistra leaves would accomplish this well.

## Tech Tips

* Dried and preserved materials accept spray paint very well. It is easy to layer shades, colors, and palettes to create a natural look or a bold look. Spray in light layers, adding additional coats as needed after each layer dries. You can always add more, but you cannot take away!

* Cutting heavy stems like these large sunflowers on an angle, and then cutting that angle a second time, creating a "V," helps lodge the stems into place within the foam. If only a single angle is cut, there is a good chance that the stems will spin once in the foam.

* By using the vessel as a riser instead of a more traditional vase, we are able to minimize the product that goes in and maximize the grandeur.

# Mod Mood

Decorative accents, when used in proper proportion, will always help elevate a composition. In this case, the addition of a painted pattern on the Midnight foam is our primary accent. The bold dye in the daisy poms, a neighbor to airbrushing, drives home the color-crazy nature of this design. Topped off with vivid dried blue button flower aerials and pink-tipped bear grass wisps, this piece will certainly put you in a Mod Mood!

## MATERIALS

- Bear grass
- Dyed daisy poms
- Freesia buds
- Small sunflowers
- Dried blue button flowers
- 12" × 4" × 4" clear glass vase
- Chevron template
- Fuchsia paint
- Black river rock
- Midnight foam

### Tech Talk

* *Repeating* the fuchsia paint on both the foam and the tips of the grass creates *unity*, thereby also creating a cohesive composition.

* The use of the round flower *forms* throughout the design creates a pleasing *pattern*.

* My tests have shown that spraying the floral foam in this manner does not affect the life of the flowers.

**1.** Place a chevron template on the side surface of a standard brick of dry Midnight floral foam and airbrush it with fuchsia paint. Remove the template and reposition for further painting as needed. The dry time is very quick. Then, soak the brick.

**2.** Following the guide marks on the brick, cut the wet foam into thirds. Using a long knife that cuts straight down and all the way through in one pass is important here, since the blocks of foam will show and you won't be able to hide any angles or inconsistencies in the cut.

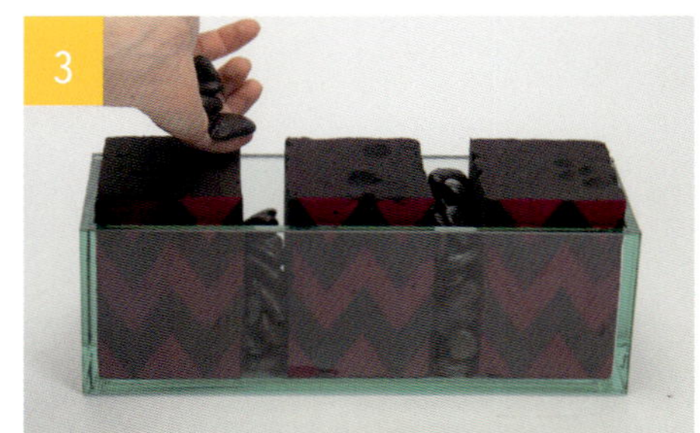

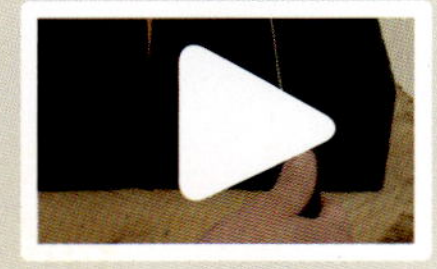

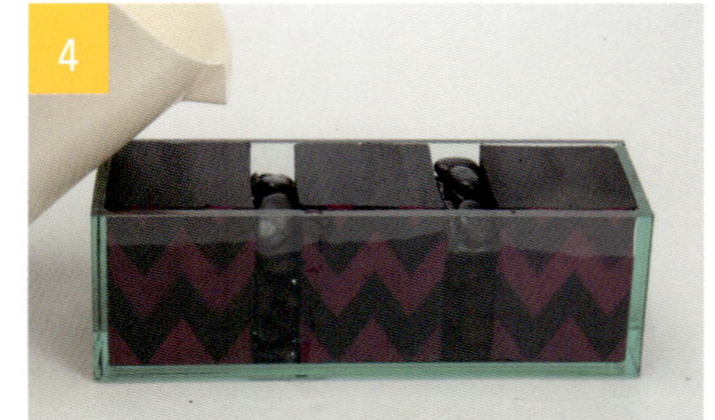

Tricks and tips on how to prepare, cut, and use floral foam properly can be found in this video.

**3.** Place the blocks into the container, evenly spaced. Fill the gaps with black river rock.

**4.** Slice across the top, cutting the foam blocks flush with the container rim and fill the container with water almost to the top. Since we're working in a clear vessel, the water line is important and will become a distraction if it is not kept high enough.

**5.** Arrange the daisies in the foam. Take care with your placement since the foam will be exposed as part of the finished design. If you must remove or move a stem, try to reuse the hole you made so that there aren't empty holes exposed. If you do need to reuse a hole, be sure to press the stem farther down than you

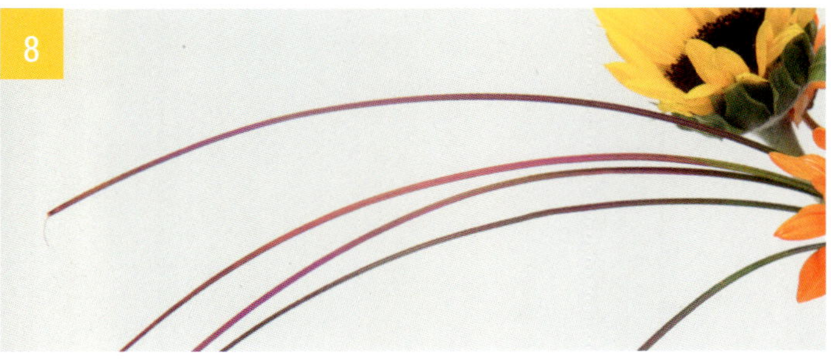

may have the first time, so that the stem has full contact with the foam on all sides.

**6.** When working with a stem that has many blooms or laterals, after considering which groups of blooms you want to use as a cluster, and running your eye from the top down the stem to locate the proper spot to cut, you can neatly dissect the stem and get multiple placements from it. Cut as close as you can to the next lateral down for a natural appearance. We don't want to see any exposed stem nubs.

**7.** Add the sunflowers and the buds from freesia stems, or another delicate stem or bud.

**8.** Airbrush the tips of the bear grass fuchsia and add these to the design in small bundles.

**9.** To complete the design, add blue buttons gathered into groupings of 3. These buttons should be left long to help balance the heavy visual weight of the dark foam bottom.

# Summary Classics

A classic design, in a classic footed urn, this is the type of piece you would likely see in a grand entrance, or banking either side of a summer wedding ceremony. To grab the eye of the viewer and to fulfill the destiny of flowers with a "face," like sunflowers and gerbera daisies, care needs to be taken that they are turned in the proper direction when placed in the design.

## MATERIALS

- Agapanthus
- Bells of Ireland
- Dried oats
- Freesia
- Gerbera daisies
- Larkspur
- Miscanthus
- Small, medium, and large sunflowers
- Snapdragon
- Variegated pittosporum
- Grande floral foam brick
- Footed urn

### Tech Talk

* This is a classic *radial design*, meant to be seen from all sides. That means your large flower faces will need to be positioned meticulously. You may also need to overcompensate for natural bends or curves in the stems themselves, placing the stem in on an atypical angle, to have the flower head end up where you need it to be.

* *Balance* is key in a design like this. Whatever is put on one side should be echoed on the other as best as possible.

**1.** Once the urn is fitted with a Grande block of wet floral foam, place the Bells of Ireland to establish the line in the design, while also establishing the height and the width. Doing this now lets you continue your work without needing to think too much about those outer parameters and the overall size goal.

**2.** As you continue to add elements like the larkspur, consider that you are creating a radial design where everything will appear as if it is radiating from one central point in the design. Angle your stems as needed to help develop this even further, and work to create symmetry on all sides. For this design, the central point of emergence, or the point of origin, starts about a third of the way up from the base of the urn.

**3.** Now place the sunflowers. With such large stems, adding them at this early point means you will have enough free real estate in the floral foam to select the best position for them. Pay attention to angling their faces for the best show and to maintain the radial design. Be sure to notch the larger stems as shown in the Sunflower Sensation project (page 34).

**4.** Place the snapdragons, agapanthus, and gerbera daisies. Be sure to sink all of your stripped and cleaned stems as deeply into the foam as possible for structural strength, and to provide the stems with the best water source.

**5.** Miscanthus, or summer grasses, are added for subtle movement and an extension of the line. Variegated pittosporum is added to fill in, create depth, and hide mechanics.

**6.** Where there are any visual holes in the composition, you can add pink freesia, small sunflowers, and dried oats.

### Tech Tips

✳ The Bells of Ireland and the snapdragons are negatively geotropic, which means they react in opposition to gravity and will send their tips shooting upward during their life cycle. If you have a bell that is curved today, and you place it in your design to accentuate that, tomorrow the tip of that bell will be shooting upward, possibly changing the line and visual path in your arrangement. Just another layer to consider when you think of the design today and beyond. See Framed by the Bells on page 148 to learn how to control the bends in your bells.

✳ It is not unusual for the Bells of Ireland to come with a frilly double leaf originating from the tips of the stems. You may decide if they help or hinder your design and can clip them off if they are distorting your desired line, or if they just don't fit in your design aesthetic.

# Birch on Birch

As the beautiful birch log decomposes, the outer layers of the bark fall off. Using these spare sheets of bark in floral compositions is very popular. In this piece, bark pieces have been layered onto a birch log to create depth and interest. Luckily, because the log also softens as it decomposes, by the time you find it and use it, the addition of mechanical elements is easy.

## MATERIALS

- Birch log
- Birch bark sheets
- Craspedia
- Eucalyptus parvifolia
- Orange roses
- Persian cress
- Pink asters
- Pink roses
- Purple scoop scabiosa
- Stonecrop
- Sunflower petals
- Staple gun
- Spool wire, 20 or 24 gauge
- 18-gauge stub wire
- Floral glue
- Black or natural-colored rope

### Tech Talk

* In this design, birch has been *layered* onto birch, and floral elements have been *layered* into the garland.

* Placement of materials in this application is up to the designer. For this example, we have chosen to *balance* each element evenly throughout, dry-placing each layer before *gluing*, to make sure the distribution is balanced.

**1.** Using a staple gun, add birch bark sheets loosely to the birch log to enhance the areas around where you anticipate your garland lying.

**2.** Cut a piece of rope or twine the length that you would like your finished garland to be. Black or natural-colored rope is recommended so that it recedes into the design. It is helpful to lay the rope on the log to visualize the finished piece and to gauge length.

**3.** To make the garland, dissect the eucalyptus parvifolia into smaller pieces or loose laterals. Bundle these cut pieces into small clusters and add to the rope with spool wire, moving farther down the rope with each bundle that you add. Do not cut the spool

wire between each addition and be sure to catch your guide rope in each bundle attachment. Not only does the rope help with length, it also provides extra support should a bundle happen to come loose. Bind the final cluster onto the rope in the opposite direction to cover the ends of the string, stems, and the wire. Drape your finished garland onto the log.

**4.** The roses that we plan to use have heads that are larger than our delicate garland would support. To help create the illusion that the roses are part of the garland without having to struggle with mechanics to actually add them into it, add a 3" piece of 18-gauge stub wire into the log where you'd like to see the roses. Keep these placements just next to the garland so that they appear attached. Use wire cutters as a hammer as needed. 1"–1½" should be hammered into the log, leaving the rest exposed for the rose.

**5.** Trim the rose and add a dab of glue to help seal the end from water loss and to help adhere the rose to the wire. Pull the wire spike from its pilot hole in the log and spear the rose's stem on the spike. Place the spiked rose back into its hole.

### Tech Tips

* Safety first! Wear gloves when piercing the roses onto the stub wire, since it's possible that the wire will come out the side of the stem and pierce right into your finger.

* The best method for removing petals from sunflowers is to resist plucking them forward, and instead to bend them backward gently. By bending them backward, they separate from the flower perfectly intact.

6. For each scoop scabiosa, trim the stem to about 1" long, apply glue generously along the stem, and adhere it to the garland.

7. Apply glue to the base of the craspedia blooms and adhere them to the garland.

8. To place each stonecrop, cut it off the plant at the soil line, then glue the stem and add it to the garland.

9. Add stems of Persian cress, sunflower petals, and pink asters, adhering them with glue.

# Metals Mix

Lucky for us, many kinds of foliage take well to airbrushing and can be used as exciting accents and in countless ways. As we see in this example, spray-painted galax leaves overlap on our vessel, creating a unique and eye-catching scalloped base, ready to be topped with a luscious arrangement. Metallics never seem to go out of trend and using them for this layered leafwork makes the composition exceptionally dynamic!

## MATERIALS

- Galax leaves
- Pink carnations
- Pink freesia
- White hydrangea
- White mini callas
- Bucket or vessel to cover
- Gold paint (Used here: Design Master's 24 Karat Gold)
- Rose gold paint (Used here: Design Master's Rose Gold)
- Pink paint (Used here: Design Master's Heather-ish)
- Floral glue

### Tech Talk

* *Airbrushed* foliage adds a layer of interest in this piece by being so unexpected. The airbrushing almost suspends the leaf in its current state. These leaves will hold up for a long time after being sprayed.

* When selecting which leaves get glued to the vessel and where, you're employing the *sequencing* technique: place the largest, and visually heaviest, at the bottom, then descend in size as you progress upward.

* The addition of the calla lilies brings a new element of *texture* to this piece. A soft or very fluffy design can benefit from the addition of a smooth or waxy-surfaced component.

**1.** Select galax leaves and organize them into graduated size groups. Airbrush, or spray-paint, the largest group of leaves with 24 Karat Gold. These will be applied to the lowest area of the vessel. Spray the medium-sized group of leaves with Rose Gold. The last layer of leaves will first be coated with the Heather-ish paint. While the Heather-ish paint is a beautiful shade of pink, it is a bit dull in its luminescence. To enhance the color and to bring it in line with the metallics of the other two paints, layer a light spray of Rose Gold overtop of the Heather-ish.

**2.** Use floral adhesive to glue the largest leaves, 24 Karat Gold, in a row along the bottom of the flower vessel. Continue adding rows, overlapping each row over the previous row. Save about a dozen of the last layer to add within the floral composition to help create unity.

**3.** Prepare your hydrangea by removing any leaves along the stems, leaving only the top leaves that will act as a collar for your design. Arrange the hydrangea within the vessel to create a rounded mound.

**4.** Add the carnations in groupings and along the same plane as the hydrangea. The desired look is a smooth transition from one flower to the next, with no depth. This can be done easier if you temporarily separate the florets of the hydrangea and thread the carnations directly through. Once you release the hydrangea, they will fall back into position, snugly around the carnation.

**5.** Next place the miniature calla lilies, also in groups of 3–4 together. Notice the new visual direction the arrangement has; with their points, the callas add line to what is otherwise a fairly lineless composition. Be sure to place your calla tips pointing in the same direction.

**6.** Place the freesia in groups, paying careful attention to the buds and how they will help the eye to move through the design. Align the buds to form a visual line leading downward to the vessel to help accentuate the round form.

**7.** Last, add into the floral area a few of the sprayed galax leaves saved from the earlier step.

### Tech Tip

Floral adhesive releases minor chemicals as it comes out of the tubes. If not given a few seconds to dissipate, those chemicals can cause leaf burn on delicate flowers and foliage, which presents as black spots on the leaf. To avoid this, when possible, apply the glue to the bucket rather than the leaf. Allow a few seconds for the vapors to evaporate, then place the leaf.

# Connections

## Banding

When creating a band around floral material, one must be
precise with their placement and portions. Banding around
botanical materials is often used to create a bold and geometric
statement, but it also serves a functional purpose by simply
holding materials in place. Whether functional or decorative,
being consistent with your banding technique, from bloom to
bloom, is key to a well curated composition.

## Binding

There are many reasons why floral materials would need to be bound together: to finish a bridal bouquet, to create a botanical structure, or to help build the perfect flower girl crown, just to name a few. Securing pieces together as you build a composition can be both beneficial and beautiful if you choose decorative binding materials that are attractive and meant to show in your finished piece.

## Staking

We all need a little support, even Mother Nature's finest specimens. Sometimes the energy goes fully into producing an exquisite bloom, and not enough gets directed to growing a strong stem. In cases like these, we can help brace the bloom by tying the stem to an upright branch or support.

# Soft & Spiky

Cactus plants, in their multitudes of varieties, and their almost-no-care-needed demeanor, are tempting for so many. What stops plant parents from collecting these interesting species are their spines. They are universally known to hurt when touched, or worse, to get stuck in your skin and be extremely hard to remove. What if a garden composed of these spiny friends had a soft and touchable accent added to it? That would be a true textural tapestry. Here, we've done just that by adding the soft yarn in distinct bands around the cacti, adding to the geometric nature of the plant forms.

## MATERIALS

- Assortment of small and medium cacti
- Aloe or other smooth-surface succulent
- Dried echinops
- Yarn in 4 complementary colors
- Fast-draining potting mix suitable for cactus
- Decorative planter
- River rocks or gravel, if needed

### *Tech Talk*

* It is not necessary to accent or *band* every cactus. In fact, unless it is done in a completely uniform manner, it may be too busy. However, certain varieties have natural depressions or curves in their bodies, and these are simply calling to be dressed up!

* The small barrel cactus in the front has been based with a yarn *braid* for even more interest.

* Adding 1 or 2 smooth-surface succulents helps balance the *texture*.

1. Create yarn balls in different sizes, from ½" to 1½" in diameter. Set aside for later.

2. If the vessel chosen does not have a drainage hole, create a base layer in the container of river rocks for internal drainage. Fill three-quarters of the way with soil.

3. While wearing gloves, set your potted cactus in the vessel. Shift, turn, and swap them until you have them arranged in a pleasing manner. To remove the cactus from the pot, grip the plant with tongs and pull the pot off. Keeping the cactus in the tongs while you fill around with soil will help guard you from the spines.

4. Using a dry paintbrush, brush any clinging soil off the cactus.

5. Wrap the yarn at various points on the plants to create bands. The varied colors and widths syncopate the design. Knot the yarn to finish, or if the plant has suitable spines, simply secure the yarn end to one.

6. Trim the dried echinop heads from their stems and position them on the soil surface. Be careful as the leaves on these stems have thorns too!

7. Finish the soil treatment with the yarn balls.

### Tech Tips

* For shorter cactus, or bands that will be positioned low and near the soil level, it is recommended to band them with the yarn prior to adding them into the garden. After the cacti are all in position and planted, you can add the remaining higher-level bands of yarn.

* Removing the yarn balls is required prior to watering to keep them from becoming unsightly. Since this composition needs extraordinarily little water, it is not a burdensome task. Once the soil top becomes dry to the touch, the yarn balls can be added back in.

* A UGlu dash is wrapped around the tips of the smooth aloe to help hold the yarn in place.

# Birch in a Bind

Who doesn't love a little bit of drama? Flowers create drama in their natural form. Their shapes, colors, and delicate and enormous sizes all work together to help shift moods and create emotions. This centerpiece is no different; it brings drama to the center of the table, with the help of the binding technique. This technique is very versatile in so many ways. Here it brings together birch branches to create a framework fit for a king or queen!

## MATERIALS

- Birch branches
- Blue hydrangea
- Lavender
- Mini green hydrangea
- Mint
- Persian cress
- White hydrangea

- Yellow iris
- 2 18" shallow design trays
- Floral foam
- Twist tool and wire ties

### Tech Talk

* This structure could also be considered a *kubari*, as it is a natural structure that is helping to support the flowers.

* One of the most exhilarating aspects of this design is employing the element of *fragrance* to help create a deeper connection to the emotion that the flowers invoke. In this design, we have employed layers of fragrance with the lavender and the mint. Mingling the two together creates a fragrance symphony.

* The *color* harmony presented here is *analogous*, including 3 hues of color that are all neighbors on the color wheel: blue, green, and yellow.

**1.** Overlap 2 birch stems that are oriented in opposite directions. Thread a twist tool wire tie around the 2 stems, and fold in half so that the round ends are parallel to each other.

**2.** Thread the hook from the twist tool into the round ends of the wire ties and pull. A few pulls by the twisting tool will crank the wires tight, creating the bind with little to no effort. At the time of printing, this version of this tool is not widely available in the US. There are larger versions that are used for twisting rebar and will do the job perfectly fine until the daintier version is more readily available. And of course, you could also bind by hand, if needed.

**3.** Continue binding on additional birch branches, orienting them in both directions, to form an "n"-shaped frame. The frame should be long enough to fit over 2 of the design trays side by side, about 36" wide. The structure should be thick enough that it creates a presence but allow generous amounts of gaps for your flowers to work through.

**4.** Set the birch cage/structure over the trays, encasing the floral foam. Midnight foam was used in this piece so that it becomes a part of the design. Read more about Midnight foam in the Stacked Steps project (page 26).

**5.** Add the hydrangea throughout the structure. Since it is a particularly thirsty flower, be sure to anchor it deeply into the foam. Keep in mind that hydrangea tend to have a natural curve to their head; take advantage of that and let the slope help round out your edges.

**6.** Place the irises throughout evenly, allowing them to stand taller than the hydrangea. This will allow their star-shaped blossoms to open and shine.

**7.** Add the lavender in groups or small bundles to give their delicate stems more impact.

**8.** Add Persian cress and accents of fresh mint to finish the design.

### Tech Tips

* The binding of the branches can also be accomplished with needlenose pliers, stub or spool wire, and some elbow grease. It is most important that the bind be done tightly so that the branches don't slide out.

* Treat your hydrangea to a dunk in a hydration solution like Quick Dip after you cut it and before you add it to the foam. This will help clean the cells in the stems so that they can drink up as much water as possible.

* If the green outer leaves are still covering the iris bloom, gently peel each side down, exposing the bloom. This will encourage the bloom to open faster.

# Orchid Dance

Orchids of all types are a timeless and classic flower that evoke feelings of elegance and grandeur. Over the years, we have seen tremendous arrangements utilizing orchids in every way imaginable. The challenge is to constantly evolve, to bring excitement to the art form of floral design, and to refine our mechanics to be second nature. The wire and midollino armature that tops this plate glass vase does all of those things and celebrates the sense of design in motion by binding green cymbidium orchids in just the right places. Without those binds, this delicate look could not be achieved.

## MATERIALS

- Blue star fern
- Green cymbidium orchids
- Oncidium orchids
- Gold 12-gauge aluminum wire
- Natural midollino
- Gold bullion wire
- UGlu strips
- 24" × 4" × 4" plate glass vase

***Tech Talk***

* The petals of the cymbidium are best shown when they are *reflexed*. By popping opening the petals as much as possible, you allow for maximum enjoyment! Learn more about this technique in the Open Orchids project (page 128).

* *Decorative mechanics* have revolutionized the floral industry over the last decade. Before the beautiful colored wires were on the scene, we would have had to complete this exercise with spool wire and then would have had to find a way to cover it. These mechanics are both functional and beautiful and certainly make designs more alluring.

**1.** To create the illusion that the midollino is banded around the vase base, apply UGlu strips in a straight line along the bottom of the vase. Neatly add the midollino, one piece at a time, until you cover the strip from top to bottom. Repeat this on all 4 sides, neatly cutting the ends so that it appears to blend from one corner to the next.

**2.** To create the spine or central armature, begin by coiling the end of the gold aluminum wire 3–4 rotations with needlenose pliers. This creates a neat, finished end and makes certain that there are no sharp edges to catch on. Then, begin to add larger loops by hand, gently twisting them and tangling them with each other. Your finished wire spine should be about 2"–3" longer than the container on both sides and will act as the support for the rest of the elements.

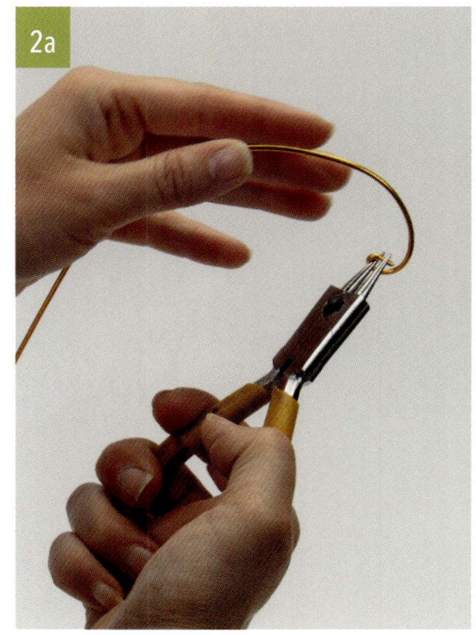

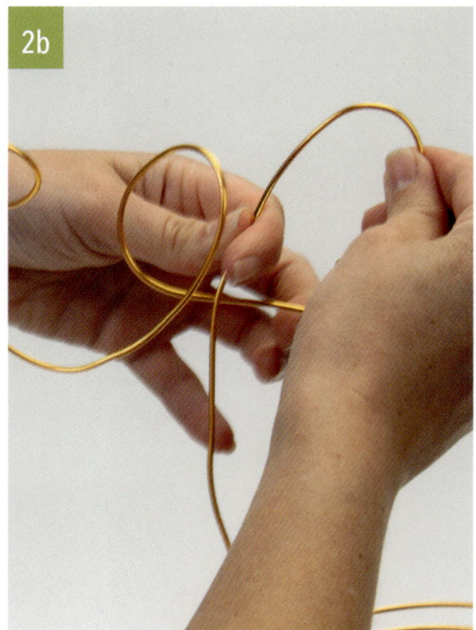

**3.** Starting with a single piece of midollino, form 2 loops and bind the center convergence point with the gold bullion. Each one of these can be slightly varied in size, some with shorter rounds and some with shorter tails.

**4.** Start to bind the midollino "bows" to the gold wire spine by fastening them in place with the gold bullion.

**5.** Rest the spine structure atop the rim of the container.

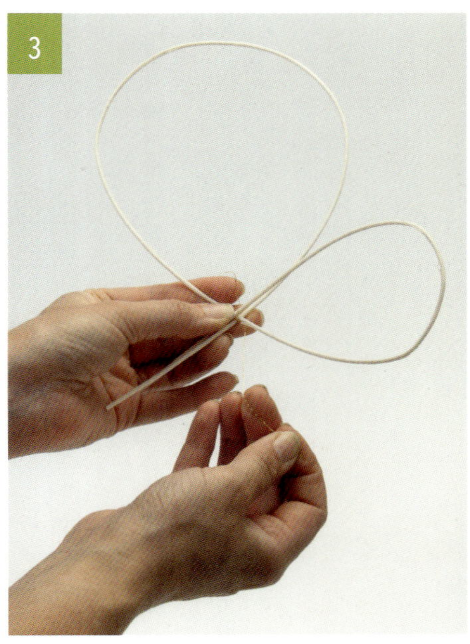

**6.** Choose an area of greater strength on the spine, where the midollino intersects with the wire, and neatly bind your reflexed cymbidium orchids to the spine. Wrap the bullion securely and just up under the head of the bloom for the strongest position.

**7.** If the oncidium orchid stems you are working with are fruitful, with many lower laterals, deconstruct the lower portions to allow for multiple placements, while still allowing the stems to feed all the way into the vase. The addition of these orchids will add to the rhythm of the structure, but take care not to overwhelm the delicate main focal point orchids.

**8.** Add a few stems of the blue star fern for contrast in color and because everything looks better with a touch of foliage in it! These leaves are shaped in a point; place them carefully, being sure to not contradict the lines that are already moving throughout the display.

7c

8

# Raise the Stakes

The techniques featured in this chapter all seem to lean on each other. One can turn into the other with a small change of execution. It is also possible that using one technique can actually mean you're using two. Such is the case in this design, where you will find staking and binding in the same action. The scabiosa and the ranunculus may need the physical support provided by the staking. But when the support is applied, in this case the raffia wrap, we're also binding the stems. If we were to increase the wraps of the raffia into concise rings, we would also be banding. Three techniques in one!

## MATERIALS

- Birch branches
- Blueberry branches
- Blue tweedia
- Buplurem
- Chartreuse reindeer moss
- Fresh blueberries
- Pussy willow
- White scabiosa
- Yellow butterfly ranunculus
- Floral foam
- Low, round ceramic bowl, approximately 15" × 7"
- Raffia wrap

### Tech Talk

* Adding the blueberries and reindeer moss presents another opportunity to explore the *basing* technique of layering.

* This arrangement is a great example of a strong vertical *line* design.

* With such a physically and visually heavy base, the branches and upward thrust needs to be substantial to make sure that the design is in proper *proportion*. Although there is transparency between the elements, the height of the branches helps balance the magnitude of the color and size of the container.

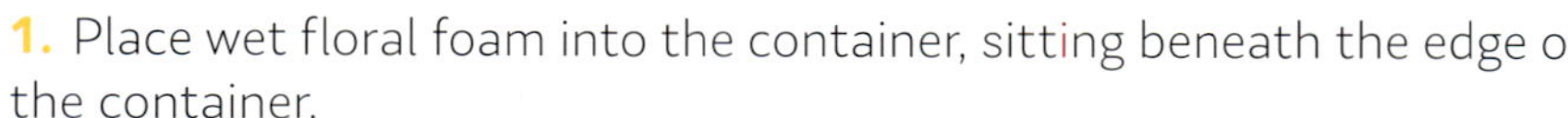

**1.** Place wet floral foam into the container, sitting beneath the edge of the container.

**2.** Create a foundation with birch and pussy willow branches. These should be placed upright and evenly around the foam.

**3.** Place the scabiosa stems into the foam, supporting them against the branches. Because the heavier heads tend to droop before the flower actually perishes, tying them into place with a small piece of raffia, just under the bloom, will help the viewer enjoy the bloom as long as possible.

**4.** Trim off extra scabiosa leaves to keep the transparency effect strong.

**5.** Add butterfly ranunculus evenly around the foam base as well. This is another flower well known for its premature droop, so use the raffia as above to tie the stems to the branches.

### Tech Tip

When tying the stems to the branches, do so with tension to hold them firmly enough in place, but lightly enough so that the stems do not suffer any crushing. Depending on the material used to tie, a single half knot is generally enough to secure it.

**6.** Now add the tweedia stems. These stems are sturdy enough on their own and do not need raffia tying.

**7.** Prior to adding the blueberry branches, edit the leaves and any berryless branches from the stems. Arrange them into the design to make a vertical focal emphasis at front/center of the composition.

**8.** Strengthen the focal area by adding shorter ranunculus to the blueberry branch area. Keeping these stems short allows for more stem stability, and they do not need to be staked.

**9.** Add a small grouping of shorter buplurem and blueberries on the left of the design to bring more visual weight to the base.

9

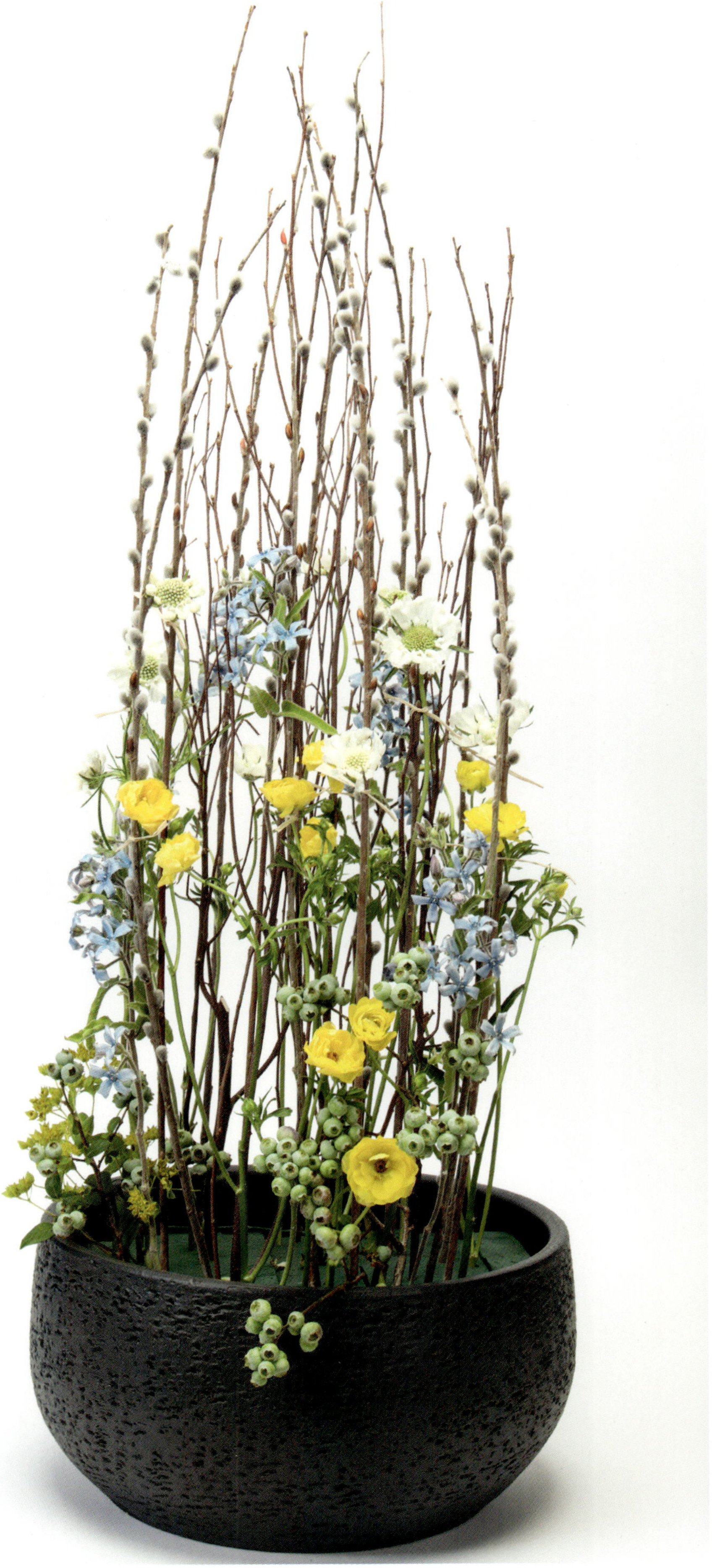

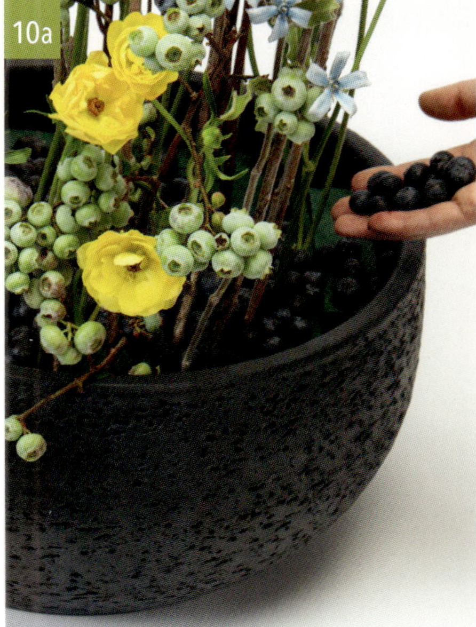

**10.** Cover the floral foam with a layer of fresh blueberries and tufts of chartreuse green reindeer moss.

10a

10b

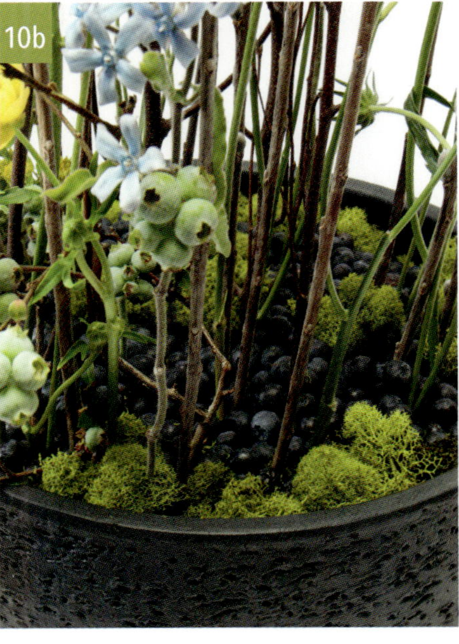

# Organized

## Collaring

To surround our compositions with repetition and a striking pattern is a remarkable way to create a finished look. Collars are usually created with foliage, as in the finishing layer to the bottom of a hand-tied bouquet, or the outer edge of a sympathy wreath. True artists who let their minds wander into new and creative territories find almost anything can act as a collar, creating that clean and finished appearance with pizzazz.

## Grouping

A simple way to assemble an impactful creation is to focus your blooms with their botanical buddies. By grouping them together, in like varieties, the viewer can fully appreciate each sampling and all of its parts.

## Hand-Tying

The hand-tied bouquet technique remains the most popular design style with brides-to-be, gals gallivanting to their proms, and even the little women who dance their hearts out on the recital stage. Creating these perfect posies takes disciplined placement of materials within the hands. A mixture of mechanics will help it all come together.

## Pavé

Imagine a quilt of the freshest flowers, where every component lived along the same plane. The surface would be coated with precious petals. The pavé style embodies this lush and level look, without allowing sections of materials to sink or create depth.

# Classic Collar

The addition of a collar around something as petite as this sweet bouquet, or as large as a grand hotel lobby piece, emphasizes everything within. It creates a tidy halo around a special collection of botanical beauties. The eye is instantly drawn in, curious to see what demands such a special treatment. It is an effective way to draw immediate interest to your composition.

## MATERIALS

- ◇ Feverfew
- ◇ Pink gladiolas
- ◇ String of pearls
- ◇ White freesia
- ◇ White kalanchoe
- ◇ Floral tape
- ◇ Ribbon and pins
- ◇ 20-gauge stub wire

### Tech Talk

* As we discussed in the Birch in a Bind project (page 64), layers of *fragrance* will always elevate a composition. In this case, it was kept earthy and sweet, with the addition of the freesia and the feverfew.

* The more string of pearls you drape over the bouquet, the stronger the *veiling* effect becomes. The *color* of this material is in contrast enough against the white that it is easily seen, and it is also light enough that it does not impede the viewer from taking in all of the details in the bouquet. The drip of the string of pearls creates a layer of drama too!

* Industry standards label ribbons by their width. The most common width of ribbon used to wrap bouquets is a #9 width, or, $^{17}\!/_{16}$" wide.

**1.** As you would prepare for any bouquet, clean all your stems by removing foliage and discarding any damaged or bruised petals or blooms.

**2.** For this bouquet, the aim is for a clean, rounded look. To help accomplish that, trim the tips off the freesia: you will use only the open blooms.

**3.** If using potted kalanchoe, simply clip the stems off at soil level.

**4.** Form the core of your bouquet with the kalanchoe by gathering them into your hands and meshing the blooms like pieces of a puzzle so that you start out with a pleasing dome. Work stems of feverfew into the kalanchoe blossoms, placing the stems not only from the sides but also threading some of them down from above.

**5.** Once there is a balanced amount of feverfew, add the freesia. Again, integrate the blossoms throughout the bouquet. Be attentive to each new flower's placement and position related to its neighbors. Bind the stems with floral tape.

**6.** To wire the gladiolia blooms, start by piercing straight through the floret with an 18" length of 20-gauge floral wire, then bend approximately 6" of the length down and wrap the parallel wires with floral tape to form a strong wire stem.

Tips on getting this tape to become tacky and take hold can be found here!

**7.** Place wired gladiola blooms around the base of the bouquet, forming a collar. Add a securing wrap of floral tape around the entire bouquet.

**8.** Trim the wires from the gladiola about halfway up the handle of the bouquet, then cover them fully with the ribbon to be sure a sharp piece of wire won't be comingled with the bare stems, resulting in a pierce to the hand of the holder.

**9.** Beginning at the bottom, wrap wide #9 ribbon up the stems. Fold under the end and use 2 pearl-tipped pins to secure.

**10.** Layer string of pearls over the bouquet, keeping in place with a dash of adhesive or wiring in the stem, if needed.

### Tech Tips

* The kalanchoe is becoming more widely available as a cut flower with some real length to its stem. However, if it is not available in your area, find a potted version that will have stems long enough to work with when the stems are cut at the soil line.

* When adding the pins to hold the ribbon in place, push them in on an upward angle so that the sharp point ends up buried within the heart of the stems and not pushed through and back out the other side.

# Magnolia Magic

For the past 20 years or so, dried flowers came with a stigma; consumers were looking for fresh blooms only. The thought of venturing into this dull, brittle, limited-variety arena seemed passé. Fortunately, we have moved into a new era where color-enhanced dried varieties are being introduced frequently, and dried and preserved floral varieties are increasing in popularity. Surprisingly popular are preserved flowers that have had all of their color bleached out of them. Not only do these varieties look interesting as they are, but they take to airbrushing very well, as discussed in the Sunflower Sensation project (page 34). This makes it easy for a designer to bring their full vision to life, without being restrained by nature's color choices. In this design, the dried and bleached flowers have been left in their cream-colored state, enhanced by the perfectly contrasted, fuzzy brown texture of the magnolia collar that surrounds them.

## MATERIALS

- Assorted dried pods
- Bleached bell cups
- Bleached bunny tails
- Bleached ruscus
- Bleached thistle
- Dried cotton
- Dried hydrangea
- Dried magnolia leaves
- Dried rosemary
- Dried scabiosa pods
- Dried strawflower
- Sheet moss
- Transparent vessel with materials to build a fitted inner liner
- Styrofoam
- Pan melt glue
- Dry floral foam
- Steel pick machine and 3" picks

*Tech Talk*

* The *texture* of the stiff but fuzzy magnolia collar is in *contrast* to the fluffy design within. This contrast helps create interest and excitement to the viewer.

* The *repetitive* nature of the magnolia around the collar also creates a *pattern* to which we are naturally drawn.

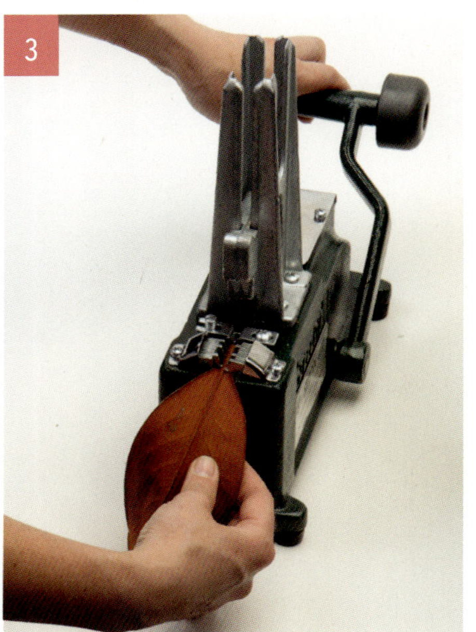

**1.** Find a single pot liner, or a combination of pieces to assemble, to fill your decorative basket. In this case, a plastic pot was the right size, albeit a bit too short. A piece of Styrofoam was glued to the bottom to bring it up to height level, and the entire inner liner was covered with sheet moss, laid on with melted pan glue.

**2.** Dry floral foam, appropriate for dried and permanent botanical designs, is set into the liner, flush with the top of container.

**3.** Using a steel pick machine, add a pick to each individual magnolia leaf. Place them into the foam, brown side up, around the perimeter of the container, being consistent with the layered direction all around the collar.

**4.** Add a few magnolia branches into the center of the arrangement, along with a few large dried hydrangea, to create the base for the design. Cut bleached ruscus into smaller segments and add these around evenly. Where the sections are too short, run them through the pick machine for extra length and to help pierce them into the foam.

**5.** Add bleached bell cups, strawflower, dried cotton, and rosemary evenly throughout. Carefully maneuver the dried hydrangea as you do this, shifting as needed, and then releasing to allow the bloom to move back into position. Stems can even be inserted into the florets of the hydrangea, but do this with care as the hydrangea is brittle in its dried state.

## Tech Tips

* When adding to this design, always keep in mind that the collar is key. All other materials should be tucked tighter in to allow the fuzzy magnolia leaves to stand out.

* For any design composed of dry or permanent botanicals, the design medium should be "dry" floral foam. It is composed of denser material and can withstand the extra pressure needed to add the stiff, wooden (and sometimes steel) picks or stems.

* When adding delicately stemmed elements like the bunny tails, it is easiest to bundle a group of 3–4 stems, wrap them with a wood or steel pick, and add into the foam as one unit.

**6.** Dried pods of assorted varieties and shapes get added next, followed by bleached thistle and dried bunny tails that have been run through the pick machine.

**7.** Finish the design with dried scabiosa, placed both down into the design and left long for aerial interest. Create unity within the design by weaving loose magnolia leaves into the wire basket frame.

# Lapel Lineup

Flowers are a lot like people; they want to bloom near their friends, but they also want to maintain their uniqueness. By grouping like elements together, not only can you create a clean and pleasing design, but it also allows you to bring proper attention to each element, as they are multiplied together.

## MATERIALS

- Cream hypericum berry
- Galax leaf
- Light-pink ranunculus
- Light-pink 12-gauge aluminum wire
- ½" floral tape
- 20-gauge stub wire
- Floral glue
- Light-brown bind wire

### Tech Talk

* The *banding* technique is used to provide professional finishing treatment to a portion of the stems of the baby's breath and ranunculus boutonnieres.

* The *wrapping* technique was used to provide professional finishing treatment to the entire stems of the green orchid and the strawflower boutonnieres.

* The *grouping* technique is often confused with the *clustering* technique. It's important to remember that in grouping, flowers retain their individual identities. In clustering, it's hard to distinguish between the individual elements.

**1.** Using needlenose pliers, turn a flat spiral at the end of the aluminum wire. Make the spiral approx. ¾" in diameter. Leave about 5" of the wire straight beneath the coil and cut from the spool.

**2.** Trim the ranunculus stem flush to the flower and remove the sepals. Flood the back of the flower head with floral adhesive. Use enough to make sure you seal the end and get a good bond, but not so much that it will squeeze out beyond the flower head when pressure is applied. Allow the glue to set for 20–30 seconds.

**3.** Press the wire coil onto the ranunculus head to secure. The wire will need several minutes to firmly take hold and bond with the glue.

**4.** Remove the lower leaves from the hypericum stem and cut the stem to be about 2" in length. Feed a 20-gauge wire through the laterals. Bend the wire down at the halfway point. Wrap the wire to the stem with a small band of floral tape. Watch the video in the Classic Collar project for tips on using the floral tape (page 82).

**5.** Layer the galax leaf with its natural stem against the wired hypericum. Use a small band of floral tape to connect them. Layer the ranunculus in front of the hypericum, again securing with a small band of floral tape. Try to keep each band of floral tape at about the same level so the mechanics to cover are minimal.

**6.** Trim off the excess stems and green wires where the tape ends, leaving only the pink aluminum wire as a tail.

**7.** Create a band of bind wire over the floral tape area. To do this, leave the first tail about an inch long, wrap neatly down and back up, then twist the second tail with the first to secure the band in place. Trim excess bind wire and carefully tuck away any sharp edges that may be exposed from the twist.

**8.** Use the needlenose pliers to curl the bottom end of the light pink wire into a spiral to create a decorative finish.

### Tech Tips

✳ As the photo shows, the stem treatments can be done in a variety of ways. Try different lengths of bindwire wrapping, ribbons, and decorative wires.

✳ Finished stem length of boutonnieres varies throughout the industry; it can be predetermined based on materials or it can be determined by your personal preference. Regardless of length, stems should always be finished in a professional and clean manner.

# Feather Flush

Bold feathers, complemented by demure orchids, make for a captivating bouquet. And a bouquet like this does not come together easily. Luckily, working in a thoughtful and restrained manner will keep you on the right track. The birch branches are used to support the more delicate bunny tails and uluhe ferns. The soft texture of the feather barbs cushions the stiff birch branches, creating comfort for the carrier. A feather cuff and orchid collar certainly make this bouquet all the sweeter!

## MATERIALS

- Birch branches
- Dried bunny tails
- Lady Amherst's pheasant feathers
- Light-pink cymbidium orchid
- Ringneck pheasant feathers, approx. 18" long
- Uluhe fiddlehead fern
- Light-pink translucent paint (used here: Design Master's Just for Flowers Pink Petunia)
- Raffia
- 1" aluminum flat wire
- 12" wired burlap ribbon
- UGlu dashes

### *Tech Talk*

* The birch branch structure created in this piece is an example of an *armature* mechanic. By building a support structure of natural, or even decorative materials, then threading your blooms through it, you can be sure that your flowers are well supported.

* The orchids encircling the binding point showcase another great example of *collaring.*

* The *binding point* is the essential crux of this, and of any hand-tied bouquet. It is the area where materials are bound together to keep the structure secure. Binds must be done firmly and with professional attention to detail.

**1.** Prepare your cymbidium blooms by removing each from the stem, reflexing the petals open as seen in the Open Orchids project (page 128), piercing the wire through the flower just below the bloom, bending the wire down, and taping as shown with the gladiolas in the Classic Collar project (page 82).

**2.** Spray the bunny tails with a light coat of translucent light-pink paint, so that they retain their fluffy nature.

**3.** Trim birch branches to an appropriate size for a hand-tied bouquet. Discard any laterals that are pointing the wrong direction, or that make the stem itself too physically or visually heavy.

**4.** Arrange the branches in your hand, meshing them into one another, creating a pleasing layout. Bind this cluster together with a short length of bind wire, or with the twist tool and wires as shown in the Birch in a Bind project (page 64).

**5.** Starting with the longest and working to the shortest, add the pheasant feathers to the birch grouping. Vary their placement within the bundle, working into the back and the front with stems angled slightly differently to add depth and drama.

**6.** Lace the uluhe fern stems into the web that is created by the birch and feathers. Choose the direction for each stem's curl to create balance. Secure the bundle with more bind wire near the base of the feathers.

**7.** The bunny tails get added in next, varying their height and placing them around the front and the back of the bouquet. Hold the stems low, near where your binding point is, to help avoid breaking these brittle stems. Wrap the bouquet with a short length of bind wire again.

**8.** Begin adding the orchids just above the binding point of the bouquet, being deliberate about how the orchids are positioned. One of the benefits of wiring the orchids is that you can slightly angle or bend the orchid heads as needed so they line up straight and with the throat pointing downward. Add a length of bind wire to hold the orchids in place.

**9.** Clip out the long wires from the orchids so they are only about 3" long. It is important that the wires are not seen or felt by the holder.

**10.** To create the feather cuff, wrap the flat wire around the base of the bouquet to measure the length of wire needed, clipping a length that allows about ¼" overlap.

**11.** Place UGlu strips at each end of the flat wire and at its center, then adhere a length of burlap wired ribbon along the flat wire. Cut the ribbon ½" longer than the flat wire on one end. Fold the extra ½" of ribbon over the sharp end of the wire.

**12.** Place UGlu strips in the center portion of the ribbon, enough dashes to cover about a third of the entire length. Press pheasant feathers into the dashes until the area is covered, leaving the feathers to extend above the wire about 4". Trim the base of the feathers away, flush with the wire.

**13.** Wrap the collar around the bouquet stems, centering the feather detail in front and with the tips pointing down. The feathers help hide the binding mechanics, create a unique and soft finish, and help extend the line of the bouquet.

12a

12b

12c

13

# Spiral Spike

The hand-tied bouquet is by far the most popular style of bouquet that is carried by brides today. We have seen it done in every type of way, and with almost every type of material. Still, brides and designers alike do not tire of this style and look. The most successful designers are leveling up their hand-ties by tapping into creative color palettes and exploring new or seldom-used materials. This bouquet, at first glance, looks to be a classically beautiful, dramatically colored, but mostly traditional bouquet. Upon closer inspection you will see that it actually contains pink grafted cactus within, placed carefully and purposefully, creating visual tension and excitement!

## MATERIALS

- Blush-pink roses
- Cream hypericum berries
- Hot-pink ranunculus
- Ninebark
- Pink grafted moon cactus
- Scabiosa pods

- White roses
- Floral glue
- Wooden plant stakes, green or blonde
- ½" waterproof tape
- Raffia

### *Tech Talk*

* By working with soft, romantic, and classic blooms, and mixing in the unexpected and bristly cactus, a great deal of visual *tension* has been developed in this bouquet.

* The *color* harmony used in this bouquet is monochromatic, including many shades of pink and accented with white and cream. When learning design and how to work with color harmonies, it can be helpful for a designer to start with monochromatic, and then introducing their color neighbors (making it analogous), and their neighbors' neighbors, and further from there. The designer will methodically learn what works together and what is less than desirable.

**1.** Prepare the stems you anticipate using in the bouquet by removing any leaves or rogue laterals from the main stems, leaving only the blooms you want to showcase. The exception to this is the ninebark, where the foliage can be allowed to extend down the stem for a short length.

**2.** Cut the cactus off at soil level. Coat the pointed end of the plant stake with glue, then skewer the stake into the green flesh of the cactus. Allow this glue to set for 10 minutes or so before handling.

**3.** Begin to assemble the bouquet in a spiral fashion, with the technique shown in the Spiral Sequence project (page 136). Begin with a large visual anchor in the center—in this case, a white rose was used—and, turning the bouquet as you build it, continue to add stems.

**4.** Once your spiral is developed and you are confident of the direction that your stems should be added, be careful not to contradict the direction at the binding point. You can thread some of the stems into the bouquet through the top, if needed, as you evaluate the composition. Make sure that the threaded stems follow the spiral pattern. Each variety should be evenly spread throughout the bouquet to create a pleasing visual balance.

**5.** Once your bouquet is mostly complete, thread the cactus into the bouquet, staying away from the outside edges so there is no chance of the bristly heads brushing up against someone.

**6.** Tape the stems, using waterproof tape to secure the bouquet.

**7.** Reach into the bundle of stems with your clippers to cut each of the plant stakes as short as possible. Then, cleanly trim the stem ends so they are as neat and flush as possible.

**8.** Using 6–8 strands of raffia, wrap and knot over the tape, then trim the ends of the raffia.

# Pavé Pattern

Most often in floristry, we are creating radial designs with levels of depth. These arrangements have angled insertions and look as though they radiate from an imaginary point of origin. A less frequently used design style is pavé. Pavé designs can have straight or angled stem insertions, but you will find that there is little to no depth or change of surface level that the flowers live on. This is a popular technique used in jewelry and it can be very chic indeed!

## MATERIALS

- Galax leaves
- Green Spanish moss
- Ninebark
- Pink waxflower
- Scabiosa pods
- White rose

- 18" shallow design tray
- 2 blocks floral foam
- Moss ribbon
- Glue gun
- 24-gauge stub wires, cut in half

***Tech Talk***

* When petals or leaves are manipulated together to appear as a new flower, a *fantasy flower* is created! The mechanics used to accomplish a fantasy flower are generally *wiring* and/or *gluing*.

* When there is little or no *depth* to a design, *texture* is of extreme importance. *Contrasting* surface qualities like soft, waxy, smooth, and fluffy all need to be considered and blended together with professional perfection.

**1.** Line the outer rim of your shallow design dish with a bead of hot glue, and wrap moss ribbon around until it is covered.

**2.** Add 2 blocks of wet floral foam; slice their tops off until they are flush with the rim of the dish.

**3.** Add the scabiosa pods in a curved "S"-shaped line, with all of their stems at the same height, about 4" long.

**4.** Next, add the white roses in groups of 2, on both sides of the scabiosa line. Keep the distance between the groups nearly equal.

**5.** To create a galax leaf rosette, sort and organize your galax leaves into size groups: small, medium, and large.

**6.** Starting with a small leaf, roll each side in toward the center until you have a cone shape. Pierce the leaf just above the stem with a piece of 24-gauge wire and fold the wires down after bending at the halfway point.

**7.** Next, layer another small, or a slightly larger, leaf around this first cone-shaped leaf. Overlap the sides as needed so that it creates a clean curl. Add a wire through both leaves and fold down. Continue in this pattern, layering larger leaves over the smaller, piercing the bundle after each leaf addition. You should have a rosette after 5–6 leaf layers.

**8.** After your rosette is formed, take one of the wire tails and wrap it up and around the other wires, twisting down about an inch to secure all of the tails around the original stems. Trim away the wires under the twist; you will put only the actual leaf stems, trimmed to about 1½" to 2" long, in the floral foam.

**9.** Place 3 galax leaf rosettes into the arrangement, nestling them into the 3 curves of the rose/scabiosa line.

**10.** Add short pieces of pink waxflower and ninebark stems, creating distinct and deliberate clusters, separating and spacing them for best composition.

**11.** Add a few dried pods for textural interest. Tuck the green Spanish moss at the rim of the container to add more texture and to help cover the foam.

### *Tech Tips*

✳ Since the overall shape of the galax leaves are round, as you roll it, there will be natural dips in the rosette petals. As you add the next leaf, layer the taller/round part of the new leaf, behind the dip in the previous leaf. This will ensure a true rosette shape.

✳ Another way to create the galax rosette is to use a stapler to staple each leaf at the base once you have it in place.

# Revealed

## Braiding

Right to middle, left to middle, right to middle, left to middle, repeat. Simple repetition of motion creates a lively pattern, and is enjoyable handwork, and the results are immediately satisfying. Botanical braids are quite unexpected, opening the doors for creativity.

## Kubari

Natural mechanics are available to us in spades through our surrounding environments. Branches, stems, roots, and more can be interwoven, banded, and bound to create a structure that supports the addition of plant material. While man-made complementing components can be helpful and add a decorative flair, their natural predecessors still reign as the go-to design style in today's eco-conscious world.

### Reflexing

Many flowers benefit from the popping of their petals. By gently inverting petals, you can create an almost fantasy-like flower, which allows for exploration of the internal structure and new and striking silhouettes. Flowers like cymbidium orchids, roses, tulips, even anthuriums, benefit from this technique, and it does not shorten the life cycle of the bloom itself.

### Sequencing

A well-organized design can bring joy to many, and having a reason and a system for how to create this organized design can be a lifeline for a budding botanical artist. Accounting for a measured shift in ascending size, quantity, color, and spacing creates a pleasing path for the beholder to journey on.

### Weaving

The constant and consistent motion of one material crisscrossing another brings a sense of organized thought and subtle tension. Excitement emerges from the addition of a weave to or around a design, most especially when that weave is the basis for an underlying form.

# Braided Bust

Creating and wearing a statement piece of floral jewelry, such as this necklace built on a woven and braided base, can set you apart from the crowd. One of the best parts about this type of piece is that it is a work of art, formed by wires, ribbons, or other materials, and requires minimal additions of blossoms to enhance it until it is perfect. After your event, you can carefully remove those blooms and the residue left behind and embellish it again for your next event. It may be a fashion faux pas to wear the same dress twice, but does the same go for fabulous floral jewelry? With the ability to redo and renew your piece, no one will ever know!

## MATERIALS

- Light-pink ranunculus
- Miniature phalaenopsis orchids
- Spurflower leaves
- Light-pink 12-gauge aluminum wire

- Chenille yarn
- UGlu dashes
- Silver bullion wire
- String of pearls
- Floral glue

***Tech Talk***

* *Gluing* is an important technique used in this design. It allows for the designer to have confidence that their blossoms will hold in place, without using unsightly mechanics.

* The element of *size* is of incredible priority here. The size of the wearer and the rest of the ensemble being worn need to be considered before making such a statement piece of jewelry. These techniques could easily be transitioned to a smaller or less layered project.

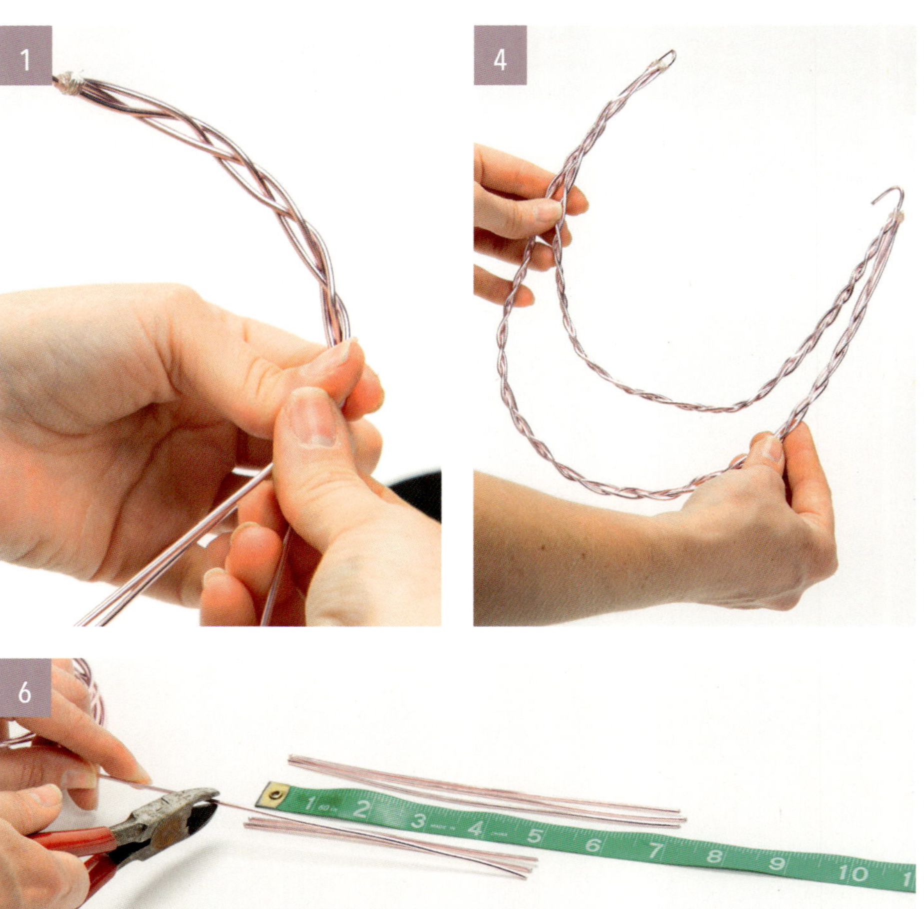

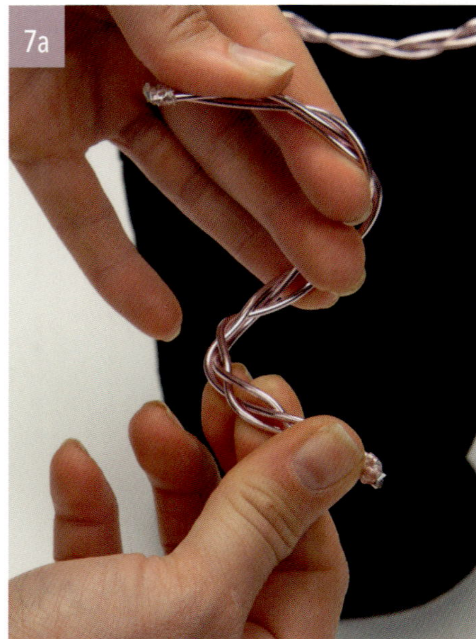

**1.** Cut 3 20" lengths of aluminum wire. Join the 3 pieces by wrapping the tips together with a third of a UGlu dash, stretching it as you wrap to get extra coverage and a thinner, less bulky application. Overlay the UGlu with a tight and tidy band of bullion wire to help hold your 3 wires in place as you braid. Loosely braid the wires, finishing the other end with the same UGlu and bullion binding technique. Your finished braid will measure approximately 19".

**2.** Next, cut 2 16" lengths of wire and 1 18" length of wire. The longer of these will be used to make the hook closure. Line the 2 16" pieces together and add the third longer piece, leaving equal length of 1" hanging from each end. Where all 3 meet, bind them with the UGlu and bullion, then loosely braid until you end on the other side, finishing with the same technique.

**3.** Using needlenose pliers, fold down one of the overhanging ends until it closes and makes a complete loop. Create a hook with the other end that fits into the closed loop.

**4.** Layer the first braided length around the second and attach with UGlu and bullion wire.

**5.** Where the connection points are, add a layer of the chenille yarn. This not only softens the physical feel of the piece and helps cover the mechanics, but it also brings in a new texture.

**6.** Follow the same procedure for creating a braid out of 5" pieces of wire and 7" pieces of wire.

**7.** Once the braids are complete, twist them into a spiral or DNA helix shape. Using UGlu to hold each end in place, place one end on the inner necklace braid and the other end on the outer necklace braid. Then wrap those joints with yarn.

**8.** Choose a few other areas along the frame of the necklace to create yarn bands so that there is even distribution of the texture. The chenille points on the main necklace serve as glue points; the flowers will be adhered to them.

**9.** Start by gluing the scented spurflower leaves to create a base. The spurflower has a pubescence to it, which helps with the bond. Stack and layer as desired to create depth.

**10.** Cut the ranunculus stems completely off the blooms and create a pool of glue on the back of the flower head as shown in the Lapel Lineup project (page 90). Allow it to set for 20–30 seconds so that it starts to get tacky and creates a stronger bond. Then place each bloom on the necklace, again adhering them to the chenille points and/or the leaves.

**11.** Cut the orchid blossoms completely from their stems and, again, use the glue-pool technique above to adhere them.

**12.** Thread the strand of string of pearls through the necklace, using a bit of glue to attach it to the base in selected spots as needed.

**13.** As a finishing touch, glue on a few ranunculus buds.

### Tech Tips

✻ Cutting the UGlu dashes into smaller pieces is very helpful when working on small projects. And since the dash can be stretched for lengthier use, it can be economically helpful as well. Since the UGlu is such a tacky material, it is recommended to cut the dash with nonstick scissors, and to dedicate these scissors strictly to this task.

✻ The UGlu is meant to temporarily hold the wire pieces together loosely, while you get a good and tight wrap around them with the bullion. In this case, the UGlu is your third hand. Keep all under wraps tidy and thin so that the yarn overlay doesn't end up bulkier than it needs to be.

# Bundled & Braided

"A little goes a long way." That saying is an oldie but a goodie and it rings true when creating impactful designs with minimal but well-organized materials. The type of design being described here is a formal linear design. It makes great use of bold forms, clean lines, and negative space. This design has not only all three of those, but a surprise braided element too!

## MATERIALS

- Ivy
- Monstera leaves
- Pink ginger
- Pink ranunculus
- ZZ plant
- Footed compote
- Floral foam
- Mirror glass pebbles
- Bullion wire

### *Tech Talk*

* Placing the second ginger just beneath the first, and the tight layering of the 2 monstera on the right, are great examples of the *shadowing* technique, as shown in the Slate Shadows project (see page 200).

* This design is composed of a variation of flower *forms*, showcasing different shapes throughout the design, without the addition of many different varieties.

* The gap between the ZZ plant stems and the sliced monstera leaves is a great example of *negative space*. While there is no physical item in this space, it is very much a part of the design and helps draw attention to the individual flowers.

### Tech Tips

* To get the rich green to come out on the monstera leaves, spray them with leaf shine or rub the leaf down with a dry paper towel.

* The monstera leaf stems have a natural curve to them, and the curves could differ from one another. When choosing your leaves, make sure the stem is bending in a way that will enhance your design and not make you work harder to overcompensate for an opposing natural bend.

* Give the ivy braid ends a sharp, clean cut for easier addition to foam, or attach the ends to a wired wood pick.

**1.** Set the wet foam into the compote, flush with the top of the container.

**2.** Starting with your ginger, double-cut your stem into a "V" as shown in the Sunflower Sensation project (page 34), and place these in a grouping on the left, sequencing them from tallest to shortest, and with a bit more of an angle, as each one goes in.

**3.** Balance the tall ginger with a grouping of the ranunculus placed middle left.

**4.** Trim the ZZ plant stems at soil level and add those into the center area.

**5.** Select 2 monstera leaves that are similar in size. Slice each lengthwise along one side of the center vein. Trim away any stray fibers remaining along the stem edge for smoothness.

**6.** Place both leaf halves in the arrangement with their curved edges forming the arrangement's left side. Add a whole monstera leaf toward the base of the arrangement on the opposite, lower right side, and a second just above.

**7.** To create the braided locks, strip the leaves from 3 ivy stems.

**8.** Wrap the ends together with bullion wire to bind them in place. Wrap the wire at a diagonal down the bundle to bring the pieces together for the full length. Then secure the other end with a bullion wrap as well. Create 3 of these bundles.

**9.** Place all 3 ivy bundles together, in line, and bind the ends with bullion wire. Proceed to braid these bundles and create one ivy stem braid; bind to finish at the other end. Create 3 of these braids.

**10.** Arc the braid from the front of the compote to the back beneath the ginger, taking care to firmly anchor the ends. Repeat with the remaining 2 braids.

**11.** Place a few ivy stems in the front left to create repetition of the materials used and to add a bit of softness and drape to the design.

# Happy Hyacinth Trails

Sometimes Mother Nature comes along and produces a perfect specimen, one that awes the flower fan with its color, form, texture, and intriguing botany. Tillandsia xerographica is that specimen! Not only does it need very little care as a common and much-loved house plant, it can also be deconstructed and added to designs to create arresting color depth and rhythm. Or, as in this case, it can be used as its own natural mechanics. Tucking into the nooks of the Xerographica allows us to enhance the natural eye candy and create a unique bouquet.

## MATERIALS

- Bleached ruscus
- Light pink hyacinth
- Tillandsia xerographica
- Translucent pink spray paint (used here: Design Master's Just for Flowers Pink Petunia)
- Bullion wire
- Floral glue

**Tech Talk**

* *Scale* is of importance when creating a bouquet like this. The Xerographica comes in many sizes; choose one that is appropriate in scale to the frame of the holder.

* The Xero itself is visually dominant, so be sure to consider *balance* when adding the hyacinth tendrils, both in regard to their quantity and to their length.

**1.** Prepare the bleached ruscus by cutting the laterals off the main stem and airbrushing them lightly with the pink paint. Let dry, flip, and apply to the other side.

**2.** Dry place the ruscus pieces throughout the nooks of the Xerographica and adjust their placement until it is balanced, using any natural curves of the stems to your advantage. Once you are happy with their placement, apply floral glue to the stem ends and add in, making sure your stem makes physical contact with the Xerographica for the best bond of the glue. For tips on gluing, see the Good Enough to Eat project (page 174).

**3.** Trim the hyacinth blooms off the stem, again dry-placing them throughout the center portion of the "Xero" until you are happy with their placement. Taking advantage of the cupped center of the hyacinth, you can nestle them inside each other to create volume and depth. Treat the hyacinth buds with glue and add in when you are happy with your composition. Again, make sure the floret is making good contact with the Xerographica.

**4.** To create the hyacinth tendrils, cut a length of bullion wire, approx. 24"–36". Scrunch one end until you create a small ball that will fit into the center of the hyacinth bloom. Thread the other end of the wire from the front/inside of the head, through the center of the flower, as if you are stringing popcorn. Pull it down the bullion until the ball nestles in the center of the bloom, holding it in place.

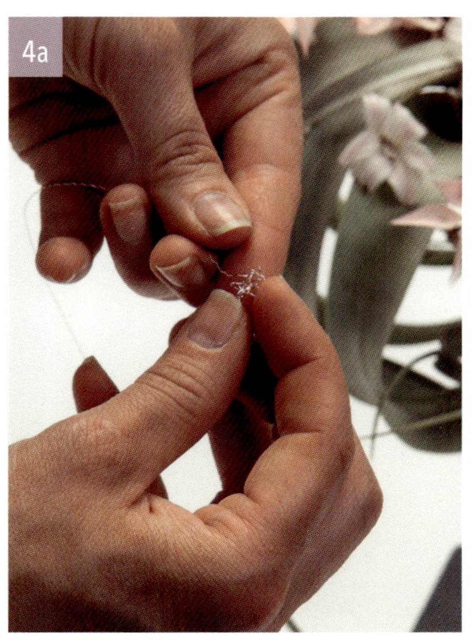

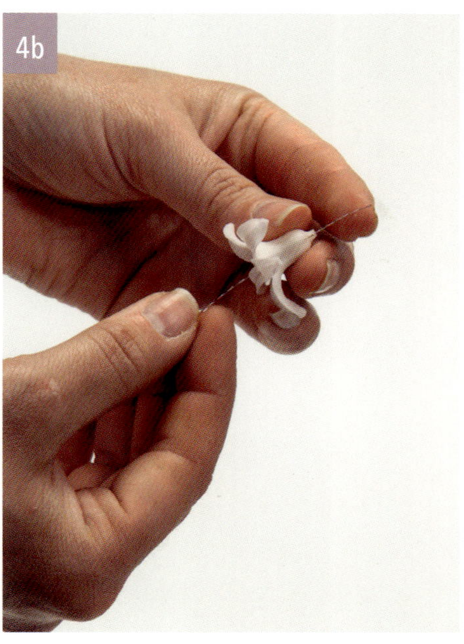

**5.** Repeat this and create about 6 tendrils, all of different lengths. You can also place several blooms on each tendril by threading the first, making a new ball slightly farther up the bullion, threading the second, and so on.

**6.** Once your tendrils are threaded with blooms, scrunch the end so that you have more surface area to work with. Cover the scrunched end with glue and add it to the inside and base layers of the Xero.

# Willow Web

If you stop and think about it, lilies are the whole package. They come with multiple blooms on each stem, making them long-lasting showstoppers. Their long stems can be the line in a design; their star-shaped heads carry an intriguing form. Many varieties carry an unmistakable fragrance. The color range that they are available in is surpassed by few other flowers. They provide their own depth by drawing the eye into the center of their blooms, and some varieties show captivating speckled patterns. If an obstacle needs to be mentioned, then it might be that their heads are naturally heavy, and therefore, the stems sometimes require support to keep them in place and upright. Creating a kubari, or a web of willow and natural support system, can do just that, in both a natural and decorative way.

## MATERIALS

- Curly willow branches
- Flat fern
- Lilies
- Penny cress
- Cylinder vase, 8" w × 14" h
- Acrylic yarn, varied shades of pink
- UGlu dashes

### *Tech Talk*

* By working with two types of different greenery here, in addition to the lilies and the branches, the *textures* in this design mesh very well.

* Lilies can sometimes get a bad rap on account of their *fragrance*. Few flowers put out such a strong scent, and many people link that smell to sad occasions such as memorials. When using a mass of lilies, be sure you know the *space* they are going to be in. Hopefully it is not confined, so the air flow helps dissipate the aroma. Or work with one of the many varieties that don't have a fragrance—the Asiatic varieties, for example, are fragrance free.

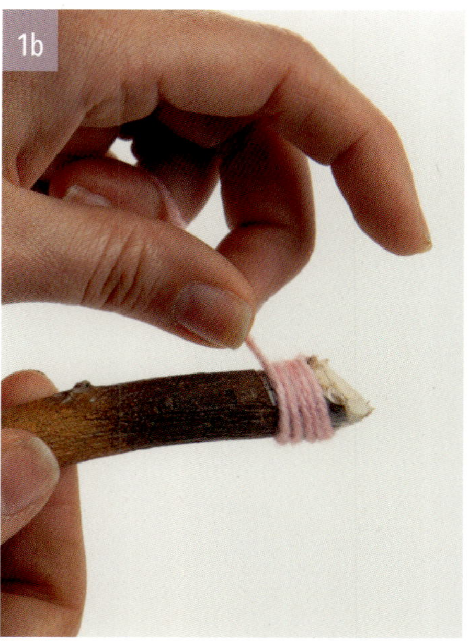

**1.** Start by cutting the thick and heavy ends of the curly willow branches in varied sizes. Embellish them by applying a UGlu dash to the end of the branch, pulling and stretching it as you wrap it around. Holding the end of the yarn down into the adhesive with your thumb, wrap the branch. To finish, tie the yarn around the branch, then tuck the end in so it is unseen. Repeat on several more branch ends, using yarn of several shades/colors and varying the lengths of the color bands.

**2.** Gather a stem or two of the thinner parts of the curly willow branches and form them into a circle. The circle should be slightly smaller than the mouth of the vase. Lower the bundle into the vase and release, allowing it to expand and be held in place with tension.

**3.** Layer the yarn-wrapped individual branches into the vase as you place the circular bundles, alternating the addition of each until you are pleased with the appearance of the kubari you have created.

**4.** Add water, keeping it below the yarn line. When you design with a shallow waterline, it's common sense but still easy to forget: take care that all stems are in the water as you add them to the design.

**5.** Arrange the lily stems into the kubari you've created. Utilize the folds and supports within the branch work to keep your stems where you place them. Make minor adjustments to the interior structure as needed. Add taller stems of curly willow, extending above the lilies to bring repetition from the bottom, and to give the height that's needed to balance such a heavy base.

**6.** Add penny cress evenly throughout to soften the look.

**7.** Add flat ferns to ground the design.

### Tech Tips

* Among the many benefits of this design is that it allows space between the lilies so that they can open freely, and each can be the star of their own show.

* The acrylic yarn will resist water, but it is still best to keep the water level lower than the yarn to keep it from getting saturated. Consider this as you place the thicker yarn-wrapped branches.

* As the lilies begin to open, if the design is going to be moved around at all, be sure to pluck the pollen-producing anthers from the center. As these puff open, they can stain the petals or anything else they touch.

# Open Orchids

Through the ages, floral designers have found ways to gently manipulate the petals of blossoms to make them stand out, while not compromising their longevity. Reflexing is one of those techniques, and it is very helpful for showcasing the full beauty of a cymbidium orchid. An orchid that has not been reflexed looks like a friend with bad posture. So be sure to open those orchids!

## MATERIALS

- Aralia leaves
- Lily grass
- Pink popcorn hydrangea
- Pink roses
- Rex begonia leaves
- White cymbidium orchids
- Pink mini calla lilies
- 8" gathering vase
- Selenite gems

- Water tubes
- Floral glue

### *Tech Talk*

* The placement of the orchids just right of center brings an emphasis to this area of the design.

* Balance and line are working so well together in this arrangement. The visual path starts with the heavy grouping of roses on the left, leading to the less heavy gathering of orchids, where the lily grass picks up and brings the eye to rest at the far right, where it's able to appreciate the distinctive calla lilies.

* The introduction of the selenite in the base of the vase is a great accent in terms of texture, color, and levity.

1. Arrange pieces of selenite to fill about 1½" of the base of the vase.

2. Lay a stemless aralia leaf wrapped in the interior of the vase, making sure the front side of the leaf is facing out. Trim the tips of the leaves where they meet the rim of the vase.

3. Remove any lower leaves from 3 stems of hydrangea and add them into the vase, resting their heads at the rim. Twist the heads into one another by gently maneuvering the laterals opened and closed. This will lock them into place and make for a smoother base in which to thread the rest of the flowers.

4. Add the roses in a group on the left by using your fingers to spread the hydrangea blooms temporarily apart, and threading through. When you let go of the hydrangea, they will bounce right back into place.

5. Add mini calla lilies, grouped together on the right, using their naturally long and sleek stems to provide line movement, leading to the center of your design.

6. Add a grouping of aralia leaves behind the roses on the left, and a few blades of lily grass extended on the right with the calla lilies.

7. Trim the orchid blooms from their main stem and place into a water tube. Coat the tube with a layer of floral glue. Make sure the glue makes

To see how easy it is to prepare and line a vase with a large leaf, check out this video.

contact with the hydrangea within the design to help hold the tubes in place. Place them in a group toward the center.

**8.** Add the rex begonia leaves in a similar manner by trimming them, with as much of the stem length as possible, from their potted parents, and adding into tubes. Layer the floral glue on these tubes as well before adding on the lower left of the composition.

**9.** Reflex the petals on each orchid by applying gentle pressure to the back of the base of the petal, while pulling the top of the petal in the opposite direction. It should pop right open!

### Tech Tips

* Because of the natural arch in the lilies' stems, it is helpful to have more on hand than your arrangement needs so that you can select the stems that naturally arch in the direction you'd like them to go. If that's not possible, employing the massaging technique, as seen in the Dripping Decadence project (page 216), can help too!

* Material that is being reflexed needs to be at room temperature or the petals will rip or snap off.

# Gradual Grass

In nature, the principle of transition happens all the time. It may not be obvious initially, but if you study the Fibonacci sequence, and how it relates to nature, you will begin to see it all around you. Using the thought process behind the gradual shift in natural materials, this design uses the sleek weaving of lily grass lengths to lead the eye from the base, up into the focal area, and straight through to the top of the bamboo.

## MATERIALS

- Curly bamboo
- Fan palm
- Israeli ruscus
- Lily grass
- Pink lily
- Plumosa

- Umbrella papyrus
- Tall and slender ceramic vase
- Floral foam
- Silver metallic wire

### Tech Talk

* Sequencing the grass ribbons helps with the *transition* from the bottom of the container all the way up into the design itself. The eye needs to make only small jumps from one ribbon to the next until it reaches the beautiful *focal* area, where the buds of the lily lead it right up the curly stem of the bamboo. This is a calm and gentle visual path, thanks to the easy transitions from element to element.

* *Harmony* between the container and the design is created here with the repetition of the horizontal ceramic bands, flowing right into the horizontal weaves of the lily grass.

**1.** Prepare the container by shaving a brick of floral foam down longways and adding it to the vessel, leaving it extended above the rim by about a half inch.

**2.** To create the lily grass weave, start by binding the tip ends of 4 blades of lily grass with metallic wire. Be sure to tuck or wrap over the initial end of the wire so there are no exposed edges.

**3.** Once you have a neat band around the tips, begin weaving between the leaves, over and under, to create a wide and flat grass "ribbon." As you finish the weave across the blades, angle the wire slightly down to start the next row below. Keep this slight angle as consistent as possible as you move down the blades. The second woven row will run in opposition to the first, where over becomes under, and under becomes over.

### Tech Tips

* There are many different types of lilies available on the cut market. Some are easier to work with than others. Some bruise easier than others. Some are more fragrant than others. A hybrid lily is recommended for this design, as they do not bruise as easily. Exploring the different lilies to learn what varieties you like to work with is encouraged.

* Cut the ends of the woven grass blade clusters sharp and neat for easier addition to the foam.

**4.** Create 3 of these ribbons of small, medium, and large length, the longest being a few inches longer than the vessel is tall to allow for an arch and some puddle on the table or surface. Add these weaves into the foam on the right, using the full side of the foam, shortest in the front, longest in the back.

**5.** Build your focal area by deconstructing a stem of lilies and reconstructing it back into the foam, one bloom or bud at a time. By doing it this way, you have greater control of exactly where each bloom lives in your design, and it allows each to be enjoyed equally. Discard any leaves that are among the blooms on the stem.

**6.** Place the bamboo stem just behind your focal area, allowing it to extend upward.

**7.** Add 2 young fan palm fronds arching around on the left.

**8.** Fill in the center area with several pieces of umbrella papyrus and Israeli ruscus, cut short and added evenly to help cover the foam. Finish with a few tufts of the plumosa for added texture.

# Spiral Sequence

Spiral hand-tied bouquets appear to be a floristry magic trick to those who haven't mastered the technique. To create a bouquet with such repetition of placement that the stems become a swirling, twirling part of the design is an art form. When done correctly and with enough stems to provide support, the stems can also act as the legs for the bouquet and hold it up, like a soldier standing in a flower cooler, ready for the battle to bring joy! Here we explore both the spiral technique and the sequencing of color, shifting from palest in the center to darkest in the outer layer of this bouquet.

## MATERIALS

- Bear grass
- Hot-pink carnations
- Light-pink spray roses
- Medium-pink spray roses
- White garden roses
- ½" waterproof tape
- ½" floral tape
- Satin ribbon
- Pearl-headed pins

### Tech Talk

* When flowers are composed in concentric rings, as they are here, whether in a bouquet or an arrangement, this is referred to as the *Biedermeier style* of design. It would take a bride who is not afraid of bold composition to carry this bouquet.

* The *repetition* of technique, strictly round flower forms, and using a monochromatic color palette brings tranquil *harmony* to this bouquet.

**1.** Prepare all of the stems you plan to use by removing any leaves from the stems and discarding any bruised or damaged petals from the roses. The spray roses should have their lower laterals removed, leaving only the top blossoms. This will prevent unnecessary bulk in the center of the bouquet and deep divots across the top surface.

**2.** Beginning with the lightest-color roses making up the center ring, hold one rose stem completely vertical and position a second across it at approximately a 45-degree angle. Grasp the 2 stems at their intersection point and rotate them a quarter turn counterclockwise toward you.

**3.** Add the third stem, angled to the left from center. Grasp the bouquet where they intersect and rotate a quarter turn counterclockwise, so the third stem is now closest to your body.

**4.** Repeat this process: place a new stem at the same angle, grasp, and rotate a quarter turn counterclockwise, adding your flowers into the bouquet in light to dark color sequence.

**5.** As you work your way through the layers, gradually lower each new flower to create a rounded shape. Work each type of flower completely around the bouquet, then move to the next color and flower in the sequence, finishing with the dark carnations.

For an array of ribbon-wrapping tips and techniques, see this video.

* Each time you grasp and turn the bouquet, it will feel like you are undoing the angle that you just put in, but don't worry—you're not. Just keep with it, and quickly you will start to see the flare at the bottom of the bouquet take shape. The flare means you're on the right track!

* When lining up the bear grass bundles, do it with care, since the thin blades will give you a paper cut–style slice very easily. To prevent cuts, let the grass sit out of water for a few hours and begin to dehydrate. Dehydration will make the grass softer and easier to manipulate.

* As you build the bouquet, keep your grip loose to be forgiving to your hand, and also to allow you to make subtle adjustments to your stems. Pulling them up or down slightly, as you go, may be necessary to make sure you achieve a clean, concentric look.

**6.** Trim the stems level and as flat as possible. The stems should be approximately 7" to 9" long and, when held at the binding point, should flare out. Narrowly wrap the stems with waterproof tape at the binding point.

**7.** Prepare the bear grass collar by binding small clusters with a narrow band of floral tape. Add those small bundles of bear grass all around the bouquet, following the same directional pattern that the bouquet stems are moving in. Wrap again with a narrow band of waterproof tape.

**8.** Retrim the stems for a level base. Wrap over the waterproof tape with a band of satin ribbon and pin in place as we did in the Bunch Bouquet project (page 16).

# Woven Wildflowers

A collection of wildflowers doesn't need much to make it feel extra special. Just the name on its own—wildflowers—makes it feel like someone took the time to go out and pick them. They must be special! Despite the already special nature of the arrangement in this project, we gave it a boost to exceptional with the addition of a lily grass woven wrap around the vase. Weaving is an art form that goes back thousands of years, but it remains impactful in today's designs because it conveys skill, persistence, and aspirations for exceptional style.

## MATERIALS

- Lily grass
- Pink lisianthus
- Pink matsumoto asters
- Pink phlox
- Pink veronica
- Plumosa
- Rosemary
- 6" × 6" clear glass cube
- Pink yarn
- Floral glue
- Craft tape
- Masking or painters' tape
- UGlu strips

### *Tech Talk*

* Although this design looks wild, thought was still given to the flower *forms* within. You will find asters as the filler flowers, lisianthus as the form flowers, veronica as the line flower, and phlox as the mass flower. When composing a recipe to design, making sure you have elements that fit these categories will set you up for success!

* The *repetition* seen in the weave creates wonderful *rhythm* to the base of the composition, which is matched equally, but differently, by the rhythm created by the flowers above it. Namely, the plumosa as it dances throughout the design and all around the perimeter, keeping the eye moving.

To see the ins and outs of this treatment, and some more inspirations, scan through to this video.

**1.** Overwrap the glass cube in a random pattern with pink yarn. This will create a surface for your weave to hold on to, and it will help create a grid across the opening of the vase to help hold your flowers in place.

**2.** A clipboard or similar sturdy and flat panel makes a convenient surface for building your woven mat. Make a masking-tape square on the clipboard, with outside edges matching the size of your desired finished square to use as a template.

**3.** Apply UGlu strips along the top of the square. These strips will temporarily hold the tops of your blades in place as you work.

**4.** Begin laying a row of vertical blades of grass within the box template, lightly pressing them into the UGlu to help hold them in place. The tips of the grass should extend up above the tape template 2"–3". Trim the bottom of the blades a few inches beneath the tape template so that your work area is easier to navigate.

We will cut them cleaner and flush to the vase in a later step. Run a length of craft tape down every other blade of vertical grass for a creative touch to your weave.

**5.** Begin to weave grass blades in horizontally, over and under the vertical blades already in place. Alternate the directions that the tips lie so that the finished weave shares the same wildflower feel that the arrangement within has. Be sure to really push each blade up and against the previous horizontal blade, keeping your weave tight.

**6.** Once the weave has filled the tape template, place a small piece of a UGlu dash on each corner, holding the ends in place. Then lift your woven square from the clipboard surface, gently pulling the vertical blades free from the UGlu strips.

7. Add a line of UGlu strips across the top and the bottom on the back of the weave to help bond it to the vase.

8. Run a thin bead of floral glue to most of the yarn on the side of the vase that you want to add your weave to. Let it sit for 10–15 seconds so that it begins to tack, then center your weave over the glass and press it firmly into the glue. Now it will be easy to see where to trim the vertical blades so that they are flush with the bottom of the vase. Once the weave is secure onto the vase, remove the temporary masking tape on the front.

**9.** Repeat these steps to create a weave for the opposite side of the vase, or the other 3 sides if you wish.

**10.** To create the arrangement, add water to the vase and start to build the foundation by adding the rosemary and plumosa. All of your stems should stay precisely where you put them thanks to the yarn wrap grid we created in the beginning.

**11.** Thread a few stems of each of the wildflowers into your foliage base: phlox, lisianthus, veronica, and asters. Where the stems have long laterals, you may be able to separate them from the main stem and get multiple placements from one stem.

# Covered

## Framing

When you've stumbled upon the most special species or done some of your most delicate design work . . . you must frame it! Botanical frames can come in the form of linear flowers or branches placed with precision to highlight your featured accent.

## Sheltering

Sheltered from the masses, it takes a keen and curious eye to see through the canopy created to the distinctive designs within. The viewer who takes the time for this treat is most deserving of the optical fest!

## Veiling

Creating an illusion of cover over a floral composition takes sheer effort. Like finding a cave behind a waterfall, such is the delight of seeing the blooms behind the botanical veil.

# Framed by the Bells

Envision with me, if you will, a bride and groom standing under a flower-covered arch. Or a grade-school goalie protecting the center of his goal as his archrival runs straight toward him. Or the precious announcement of a newborn, set in a shadow box with an imprint of the cherub's tiny fingers and toes. What do all these images have in common? They are all framed by a surrounding border. The border establishes boundaries for the setting and directs the viewer to appreciate what is within. We can accomplish this with our designs using many, many different types of materials. In this design, by working against gravity, we can manipulate the Bells of Ireland to create a frame that is as dramatic, or as soft, as we want it to be.

## MATERIALS

- Bells of Ireland
- Croton plant or leaves
- Flocked birch or yellow blooming branch
- Israeli ruscus
- Orange gerbera daisy
- Yellow pincushion protea
- Footed compote
- Floral foam

***Tech Talk***

- The color scheme seen in this design is *analogous*, meaning it is composed of 3 hues that are beside one another on the color wheel.

- The speckles on the croton leaves create one of nature's finest *patterns*.

- A *layering* effect is happening where the croton leaves are placed one in front of the other, with no space between them.

**1.** Place the wet foam in the compote, cut flush to be even with the rim.

**2.** Add the Bells of Ireland on each side of the foam, creating your frame. Stagger the heights so there is varied interest.

**3.** Create a focal point in the lower center area of the design with the pincushion proteas, placing them in a group with varied depth.

**4.** Strengthen the focal point and add color saturation by adding orange gerbera daisies.

**5.** Clip larger croton leaves from their stems and add to the foam in a layered fashion behind the gerberas and in front of the pincushion.

**6.** Add the birch branches to strengthen the framed lines of the bells.

**7.** Israeli ruscus can be cut into short sections and placed into the foam for depth and to help cover mechanics.

### *Tech Tips*

✳ As with any design that uses floral foam as its base, make sure to top off the reservoir or vessel with water when you're finished designing. This means you may have to brush a leaf or two aside, usually in the back, to see where the water level is. The alternative to this technique is to hook your finger over the side and add the water until your finger can feel it reach the top.

✳ Since we are using the croton leaves low and against the foam, the stems should be strong enough to slide into the foam. If this is not the case and it feels as though the leaf needs extra support, attach it to a wood pick and then add to the foam.

✳ To force the Bells of Ireland to have a curved tip, lay them as flat as possible while they are drinking water. In a matter of hours, they will start to fulfill their negative geotropic destinies and will curve upward. By laying a tuft of bubble wrap under the blooms, you will keep them out of the water so that the blooms don't deteriorate. See more about this fact in the Summer Classics project (page 44).

# Red Lava

This design is as much about the glass vessel and the accents within as it is about everyone's favorite flower: the red rose. The red rose will always be the classic symbol of love and drama. Here we are expressing that drama framed under two perfectly angled monstera leaves nestled within gritty lava rock and shiny mirror glass. As the observer takes in all angles and sides, they are treated to new and different views.

## MATERIALS

◇ Chartreuse reindeer moss

◇ Monstera leaves

◇ Red roses

◇ 12" round bowl with angled opening

◇ Midnight foam

◇ Pan melt glue

◇ Lava rock

◇ Mirror glass

### *Tech Talk*

✱ This technique could also be considered *sheltering*, since the monstera leaves hover over the roses. However, because there is so much distance between the leaves and the roses, the framing technique is presenting more strongly.

✱ This composition is an interesting study on *form*. The leaf canopies, the roses, the bowl, and even most of the lava rocks are round; however, the shape of the frame rising out of the bowl makes it feel equally rectangular, which is fascinating because this space is primarily void of material. So the competing rectangular form is predominantly negative space that feels very physically present.

1. Turn on your preferred nonstick skillet or glue pillow pan and get the pillows melted.

2. Trim the Midnight floral foam into a block about 5" × 5" × 3" and bevel the top corners.

Dip the bottom half of the floral foam in the hot glue. Use a spatula or paint stick to spread it evenly over the surface, being sure to cover any gaps. The glue will seal the foam and hold the water in.

3. Soak the floral foam in water.

4. Place the foam into the base of the Bias bowl. If your sealant glue is still a bit tacky and grips onto the glass, that will add stability—a bonus!

5. Choose 2 monstera leaves with similar leaf canopy sizes and opposing stem angles. Place the monstera leaves to frame the block of foam, leaving one taller and overlapping the other.

**6.** Place the lava rocks around the foam. Fill the gaps between the lava rocks with the mirror glass.

**7.** Add 3 roses to the foam, low and with complementing angles. If the roses are warm enough, you may choose to reflex one or all of the roses by popping their outer petals open, as shown in the Wrapped Wreath project (page 210).

**8.** You may also like to try removing the center petals of a mostly opened rose with a quick pinch. Seeds in the center will now show, reminiscent of a true rose from the garden.

**9.** Finish the design by adding some tufts of the chartreuse-green reindeer moss among the rock border to complement the shade of green above.

Quick tips on how to make those large leaves look their best and level up your design can be found in this video.

# Sheltered Sanctuary

A bit of mystery is always a bonus. Sheltering is the floral-design version of a great mystery novel. It leaves the viewer searching for more, wondering what is under there, how did they do that? Usually as designers we want every bloom to show its full splendor, with no obstruction. Sometimes, though, a light sheath over the design can add depth, color, and rhythm.

## MATERIALS

- Green trick dianthus
- Orange pincushion protea
- Ornamental pineapples
- Red hypericum
- Red ti leaves
- Variegated flax
- Low trough-style ceramic container
- Chicken wire

### Tech Talk

* The visual *textures* in this trough could be described as fuzzy (dianthus), waxy (hypericum), smooth (ti leaves), sleek (flax leaves), and prickly (protea). One touch of the protea will expose another mystery: they are soft to the touch!

* Consider the angles that you are creating with your flax leaves, since these are the primary *line* in the design. They do not all need to be physically touching, but they should visually link to each other without a big valley for the eye to cross.

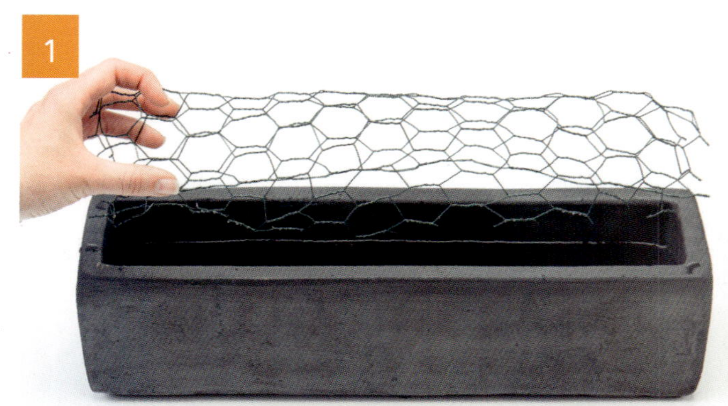

To take the mystery out of cutting and forming chicken wire, watch this video.

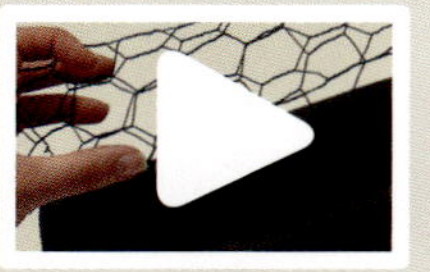

### Tech Tips

* If you have never partially cut a vein out of the back of a leaf, it is recommended to practice a few times until you get the feel of it. Also, when you are pulling the knife toward your body, please do so with care. Safety first!

* If you find that the full ti leaf is making too large of a roll, you can split the leaf in half, straight down the center, with the tip of your knife on the table. See how we did this with the monstera leaves in the Bundled & Braided project (page 114).

**1.** Prepare the trough for flowers by adding water and setting a tube of chicken wire nestled within.

**2.** To roll the ti leaves, place the leaf upside down on the table. About halfway down the stem, where the vein begins to get thickest, slide a knife halfway into the vein. Pull the knife all the way down, removing the back half of the vein. The leaf will still be able to hold its structure and take up water, but it is now more flexible.

**3.** Roll the tip of the leaf until it meets back with the leaf body and staple it in place. Repeat to make several rolls, then arrange them in the trough.

**4.** Add flax leaves, at varied heights, and leave upright, for bending later in the process.

**5.** Place the protea throughout the trough to create a base.

**6.** Remove the large leaves below the hypericum berries and add those into the design, along with the green trick dianthus.

**7.** Trim the leaves from atop the ornamental pineapple. Add the individual leaves throughout the composition, keeping aware of the sharp serrated edges on the pineapple leaves.

**8.** To create the sheltering effect, sharply crease the flax leaves and bend downward.

**9.** Some of the flax leaves may benefit from sliding into another, which will hold them in place with tension. To do this, simply slice your knife into the leaf to create a slit, then thread the other leaf into it.

# Grass under Glass

Why do something only once, when twice can be just as nice? Here we are sheltering our miniature bark pot design with twisted and tamed wired wool. Then, a glass cloche lowers over the base for an all-embracing shelter.

## MATERIALS

- Flat of oat grass
- Mini carnation
- Orange dubai Ornithogalum
- Orange spray roses
- Yellow freesia
- 15" glass cloche
- Long serrated knife
- Miniature container
- Floral foam
- Styrofoam
- Orange wired wool
- Red 12-gauge aluminum wire
- Gold bullion wire
- Burlap ribbon
- Pins

### Tech Talk

* This composition has a sense of *opposition* to it. The gritty, organic soil visually contrasts with the clean and sleek glass.

* The bark pot design being sheltered is clearly of *emphasis* in this composition. This type of look is great for highlighting heirloom or personalized containers.

**1.** To cut a round piece of oat grass, pull the flat out of the tray and work from the underside. Find a circular template that is slightly smaller in diameter than your glass cloche base, roughly ½" smaller. A plastic food container lid was the perfect size for this template. Trace the circle on the base of the grass block, using black permanent marker.

**2.** Use a sharp blade like a florist's knife to start your cut. Then use a long serrated knife to saw out the circle. Lift the circle from the block and place it on your base surface. For this look, a log slice or a cake stand makes a great base.

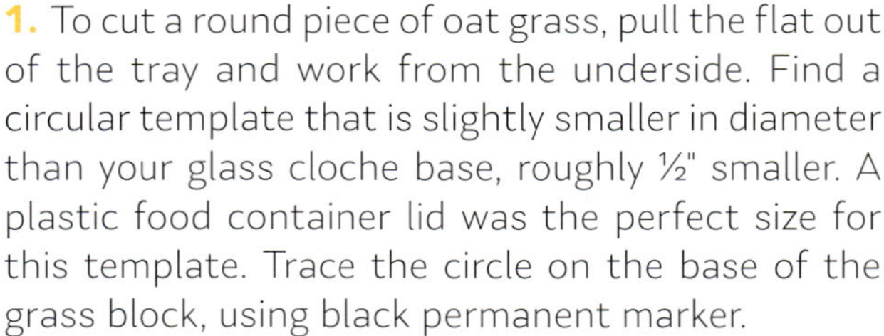

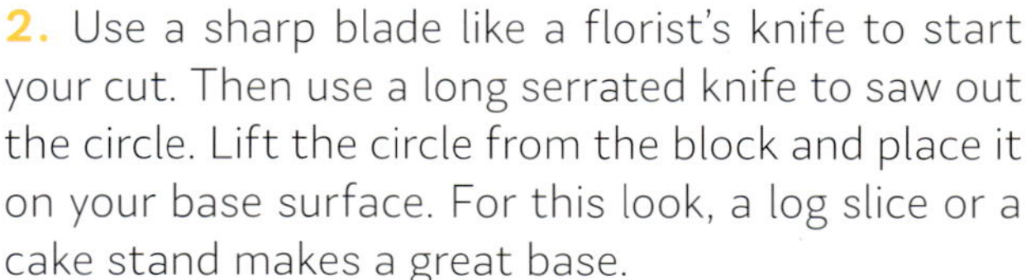

**3.** Measure the circumference of the grass circle and cut a length of wired burlap ribbon to cover the bottommost portion of the soil layer. Pin it in place, folding the end under with sharp creases of the wire for a clean finish.

**4.** Nestle a small block of Styrofoam in the center of the grass circle to serve as a boost for the small container. Fill the small container with wet floral foam, ensuring the foam is below the rim.

**5.** Place the small container in position on top of the Styrofoam and grass to make sure it nestles nicely. It may take a bit of downward pressure or adjusting to get it level and settled. When you're comfortable with the position, remove the container to a steady surface to work on; it will be easier to arrange the flowers that way.

**6.** Place 2 stems of the orange dubai low and in the center, then insert spray rose blooms that have been clipped from the main stem evenly around. Add a few stems of freesia.

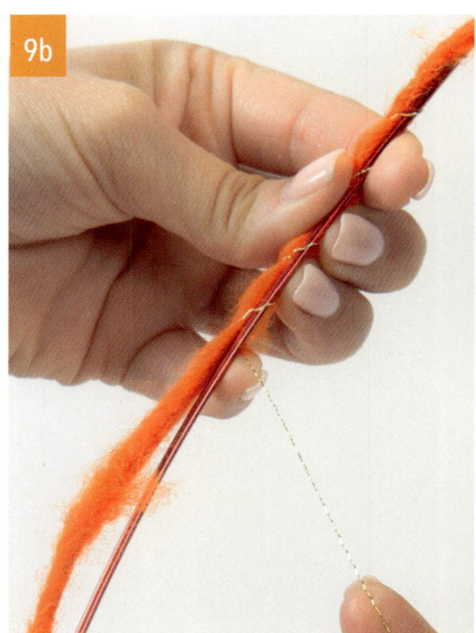

**7.** For an interesting and unexpected element, grip the petals from a mini carnation bloom and pull firmly, removing all of the petals. The shell that is left can be added to the design in small groups, in place of an accent filler.

**8.** Once the arrangement is full and with no gaps, place on the Styrofoam in the center of the grass.

**9.** To create the wire tendrils, follow steps similar to those shown in the K9 Collar project (page 178). Start by cutting a length of wired wool. Cut a matching length of red aluminum wire. Lay the red wire parallel against the wool and begin to wrap one of the ends tightly with the bullion wire both to bind them together and to

keep the wool from slipping down and exposing the internal wire. Without cutting the bullion, begin to wrap down the wool and wire, binding them as one. When you reach the other end, create a band of bullion to finish it off. Attach a wired wood pick to both ends and then begin to twist and turn the wire, creating a rhythmic shelter. As you are creating these, eyeball them carefully to make sure they are not taller than the cloche will allow. Secure the dancing wire into the soil.

**10.** Repeat the wired-wool technique 2 more times, creating 3 wire tendrils. Play with your arrangement, perhaps leaving 1 or 2 tendril ends coiling up into the air.

**11.** Place the cloche over your masterpiece.

### *Tech Tips*

* When preparing the freesia to go into the miniature design, it can be helpful to remove most of the larger, lower blossoms to keep them from overpowering the composition.

* The grass and the flowers will transpire while under the glass, causing it to fog up. A trick to stop that is to keep a small item wedged under the glass to allow the air to escape.

* As the grass grows, you can trim it to any height with a simple pair of scissors and a steady hand.

# Tamed Flames

You've heard the old saying "where there's smoke, there's fire." What if we explored smoke and fiery flame-shaped flowers in an interpretive design? Well, we've done just that here! The heliconia and psittacorum "flames" are reaching high up into the air, mingling with Spanish moss "smoke." The smoke is falling down and around the flames, creating a veiling effect. The viewer has a pretty good idea of what's happening under that smoke, but as we discussed in the Sheltered Sanctuary project, a little mystery and intrigue in floral design can be a good thing!

## MATERIALS

- Anthurium leaves
- Heliconia
- Psittacorum
- Spanish moss
- 8" × 8" × 8" plate glass cube
- Floral foam
- Black river rock

### Tech Talk

* The *size* of this piece is to be considered when planning the display area. The combination of the large *form* of the heliconia, and their bold *color*, can be overwhelming. This would be something appropriate for a large corporate office or hotel lobby, an entry table in a grand foyer, or on either side of a bride and groom as they take their vows in a tropical ceremony.

* The plate glass cube is thick and visually heavy; therefore, it needs a strong, complementing design for *balance*, and keeping the design rising from the center of the cube strengthens the rectangular *form*.

**1.** Line the cube vase with anthurium leaves that have had their stems removed, curving them against the glass as needed, with the front side of the leaf showing through the clear glass. Overlap leaves as necessary to cover natural gaps or openings between the leaves. Fill the interior of the cube with wet floral foam. Start with one large piece, then add additional pieces as needed, to fill the space.

**2.** Add water, making sure the water level is almost to the top so that the line of the water does not compete with the geometric pattern on the submerged leaves.

**3.** Trim the tops of the anthurium leaves off to be flush with the rim of the cube.

**4.** Start by adding the heliconia first, since it has the largest stem. This will ensure they get their proper real estate in the foam. Double-cut the stems so that they have a "V" shape on the stem bottom.

**5.** Add the psittacorum stems, confining those stems in and around the heliconia, maintaining the strong columnar shape of the design that is echoing the form of the cube vase. As you arrange, be mindful of the stems' thicknesses and how they vary. Try to restrain from placing these chunky stems too closely to one another, as it will cause a compromise, or a crumbling, of the foam.

### Tech Tip

Four large leaves were available when this design was created, so it took only one leaf per side to line it. However, if only smaller leaves are available, overlap them, taking care to line up the overlaps evenly all around. This design would look equally striking with pandanus layered and lining the vase in a vertical direction.

6. Take a handful of the Spanish moss and pull it to separate, without completely ripping it apart, so that it becomes thinner and more transparent. Drape the moss over the flowers, wedging it between the tips and letting it drip down the outside.

7. Cover any exposed foam with black river rock.

# Collaborate

## Gluing

Adhesive makes a lot of floral artistry possible: from large-scale seed- and petal-covered parade installations to miniature ring bearer boutonnieres and everything in between. When the handling instructions are followed, glue allows the designer to create with full confidence in their creations.

## Lacing

By intertwining stems into each other, a simple but strong foundation can be built. A strong and stable support is essential for everyday floristry as the designs are handled and delivered.

## Shadowing

A reflection of plant material can be special and calls for contemplation of placement and material choice. By layering two alike stems, one just beneath the other, a mirrored effect unfolds, leading to depth and dimension.

## Tailoring

Like a fine suit, flowers and foliage benefit from a good dose of tailoring as a way to reveal their best selves or attain a new look that may not have been seen before. By utilizing mechanical modifications, such as wires, glue, staples, and pins, petals and leaves can be manipulated to soften, harden, round out, or elongate the essence of the elements.

## Massaging

With careful, miniature manipulations, we can meld and mold many different types of materials with the warmth and pressure from our experienced hands. With practice, the amount of force and mechanical movements needed to create these bends and curves becomes second nature, and they can quickly be shaped to expose their inner excellence.

# Good Enough to Eat

Working fruits and vegetables into floral designs is very popular. It can be unexpected and exciting for the viewer. The two truly go hand in hand in nature: all the more reason to put them together in our designs. Often the fruits are cut into wedges or halves. Sometimes they are used as accents to surround a composition. Here, we have an Italian eggplant as the base for our design, and we are able to utilize the natural dip of the stem area as a platform for gluing.

## MATERIALS

- ◇ Elaeagnus
- ◇ Eucalyptus parvifolia
- ◇ Green amaranthus
- ◇ Italian eggplant
- ◇ Ivy
- ◇ Sheet moss
- ◇ White lisianthus
- ◇ White spray roses
- ◇ Floral glue

### *Tech Talk*

* The second piece of moss *layered* on top in the beginning steps of this design should be *pillowed* to allow for volume to work into.

* Adding the floral glue to the bottom of fresh stems not only helps them adhere to other stems or surfaces but, as an antitranspirant, also seals the ends, providing one less escape route for the moisture within the blossom and stem.

**1.** Coat the back of a fitted pad of sheet moss with a layer of floral glue and press it into the stem area of the eggplant. Add a second piece of moss, glued onto the first piece, with its layers pulled up and separated a bit, to allow for stems to work into this natural cushion.

**2.** Trim the smaller laterals off the main eucalyptus stem. Coat the bottom ½"–1" of the stems with floral glue and push these into the moss cushion in several places around the eggplant. The natural bend in the tip helps create the movement in your design. Gluing into the moss cushion provides many contact points, making the stem's attachment secure.

**3.** Dissect the hanging green amaranthus into smaller pieces, looking at each section carefully and assessing how it can provide the best depth and drape to the design. Strip the bottom inch of the amaranthus stem clean, apply a coat of glue to the stem, and push into the moss cushion. Repeat this technique with the elaeagnus.

**4.** Cut the buds and blooms of the lisianthus off their stem at lengths varying from 3" to 5", depending on where you plan to place it in the moss cushion. Coat stems with glue and add inro the composition. Repeat this process with the spray roses.

**5.** Finish the composition by gluing in some sprigs of ivy to bring in the dark-green color and to fill in the overall shape.

**6.** This technique works on squash and pumpkins as well and is the perfect centerpiece for a harvest-designed tablescape.

### Tech Tips

* To adapt this into a longer-lasting design, hollow out the eggplant, line with poly foil and foam, then add flowers as you would a typical design.

* The amount of time to allow the glue to get tacky should be adjusted based on the amount of glue used. A thin layer on a stem can begin to tack in 10–15 seconds. A thicker puddle could take 20–30 seconds.

* The longevity of this design is limited, as there is no water source for the flowers. However, by choosing your products carefully and treating them with professional care from delivery to design, you will get many hours out of this composition, making it perfectly suitable for a soiree.

# K9 Collar

Wearing flowers is a treat. It instantly elevates your outfit, your demeanor, and your spirits! Involving our pets in weddings and special events creates an opportunity to design an ensemble for them to wear too. Designers participating in styled shoots worldwide have created works of art for species ranging from guinea pigs to horses. The main technique to creating these mini—or in the case of the horses, big—masterpieces is gluing. Beneath the flowers lies a supporting structure, usually made of wire and then enhanced with layers of highly detailed botanical goodies. That combination of techniques is precisely how this next project comes together, modeled perfectly by Bella the Chihuahua!

## MATERIALS

- ◇ Dried white limonium
- ◇ Italian ruscus
- ◇ Plumosa
- ◇ White button pom chrysanthemums
- ◇ Rustic wire
- ◇ Bullion wire

### Tech Talk

* Adding the bullion along the rustic wire not only helps with the structure of the base but also adds a bit of shimmery *texture* to the rough rustic wire.

* *Scale* is of utmost importance when working with flowers to wear. It is helpful to obtain the length needed for a human crown or a K9 collar, so that the pieces are neither too large nor too small. Additionally, consider the overall appearance of the wearer. For example, a hairless Chinese crested dog would look smashing in a very sleek, slim, and refined collar, while a fluffy golden retriever would need more umph and body to their collar.

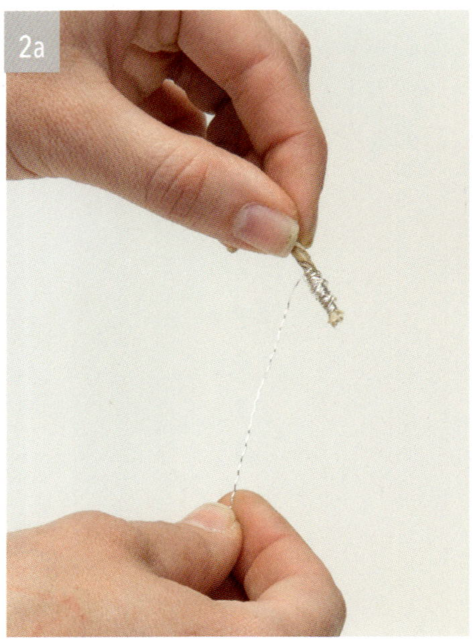

**1.** Cut a 2-yard length of rustic wire.

**2.** To make the collar base, wrap a tight band of bullion around one end. Without cutting it from the spool, wind the bullion down the length of the rustic wire until you get to the other end. Finish with a tight band of bullion on the other end and cut free from spool. This technique will keep the rustic overwrap on the wire from slipping down, exposing the structural wire underneath.

**3.** Starting 3" from one end, make a loop approximately 1" in diameter and twist twice to hold it in place. Repeat every inch along the wire, leaving a 3" tail on the other end.

**4.** To add the greenery, trim off the thick stem ends of two ruscus branches, leaving only the thinner and flexible portions to work with.

**5.** Starting at one end, weave the ruscus through the chain. Wrap it through and around the loops. The goal is to get it nice and tangled within the wire. Once that piece is in place, run the second piece through, starting at the other end. The tips of the ruscus should be on the ends of the collars so that the blunt and cut stem ends get buried in the middle. Manipulate any long or loose laterals of the ruscus to weave into the base to ensure coverage of most of the loops.

### *Tech Tips*

* To keep your work area neat and your hands clean, when opening a new tube of glue, use needlenose pliers to clamp the tube nozzle almost closed, allowing only a slow bead to come out when the tube is compressed.

* Try to avoid putting so much glue on the back of the poms that when you press it into place, it spills out. Learning how to meter the glue takes some practice but is easy to learn.

* When choosing elements for an animal, or even small children, choose hardier products like dried materials or chrysanthemums as we've used here. If the collar or crown is roughly removed by the wearer, it's quite possible it can be saved and put back on if you have strong mechanics and you use sturdy products.

**6.** Repeat this technique with 2 pieces of plumosa.

**7.** Cut the dried limonium into small pieces. Add a layer of glue to the base of the stems and add into the folds of the base created by the wire and the greenery. Be sure the glue makes good contact with the elements in the base.

**8.** Trim the larger open white button poms completely free from the stems, leaving a flat back. Puddle a small amount of glue on the back of the pom and push into the base. Add these blooms in all around, deeper within the collar, allowing the glue to set for 20–30 seconds before adding it into the base for a faster bond.

**9.** Trim the smaller, budded poms from the stem, leaving about ½"–1" in stem length. Add these all around to create depth within the collar.

**10.** Once the glue has been allowed to fully dry, wrap it around your best friend's neck and hook the two ends together to keep it in place.

### Tech Tips

❋ The floral glue bonds only to porous surfaces, so flower stems and textured materials are perfect to work with.

# Crafted Cone

How can something be both larger than life and delicate and demure? The answer: by crafting an oversized vessel and filling it with fragile and pale materials. This wall hanging is a perfect example of that. The cone itself measures 3 feet long, wrapped in a combination of vines and mosses, making it a sizeable piece. The dried-flower design emerging from the top of the cone has a well-balanced base at the point that it emerges from the cone but is finished with light and airy components, making it both large and light. The willow armature that the dried stems are laced through helps keep all of the materials in perfect position.

## MATERIALS

- Angel vine garland
- Bleached bell cups
- Bleached bunny tails
- Bleached lunaria
- Curly willow
- Dried amaranthus
- Dried angel breath
- Dried limonium
- Dried magnolia
- Lady Amherst's pheasant feathers
- Plume grass
- Preserved eucalyptus
- Sheet moss
- Spanish moss
- Foam flowers
- Floral tape
- Chicken wire
- 4 ⅛" steel bars
- Bind wire
- Rustic wire
- Cable tie
- Floral glue
- A form to wrap the chicken wire around

***Tech Talk***

✱ This project includes *gluing* of the moss, *binding* of the vine to the mesh, *layering* the form with the mosses and vines, *grouping* of several of the materials, *bunching* of the bunny tails, *hand-tying* of the curly willow, the creation of a *kubari* as described in the Willow Web project (page 124), and *lacing* of the materials through the kubari.

**1.** To create the cone, wrap a 2-foot length of chicken wire around a construction cone or other form, starting with the tip. Then, layer a second piece of wire around the top portion of the cone or form.

**2.** Using needlenose pliers, twist the loose ends of the chicken wire together to hold the form in shape. Create a hanging hook for the back of the cone using the rustic wire, wrapping it tightly around the mesh with needlenose pliers.

**3.** Thread 4 steel rods through the mesh, weaving it every few rows as you go to create structural stability.

**4.** Bind the ends of the steel rods together with a cable tie where they meet at the tip of the cone.

**5.** Starting at the bottom, attach the angel vine garland to the mesh with bind wire, then spiral-wrap the vine up the cone to cover it, leaving gaps between the wrap that will be filled in with moss. Attach the top end with bind wire as well.

**6.** Pull the Spanish moss into garland-like strips and wrap them in the gaps between the angel vine. Pressing it into the angel vine will tangle the moss into place.

**7.** Place tufts of sheet moss here and there on the cone to cover any remaining open surfaces and to add texture and color interest. Coat the back of the moss with floral glue and press onto the form.

**8.** Weave curly willow stems in through the structure and hold them in place with discreet bind wire connections that thread through the chicken wire and back out.

**9.** To create the willow structure that the dried materials will be laced through, start with a stem of curly willow, bend the tip down and back onto itself, and hold it in place with bind wire. Repeat this technique until you have about 10 different curly willow loops. Interlace the loops to create a kubari structure. Take any free laterals and tack them down with bind wire horizontally so that you have a rounded collection to work with.

### Tech Tips

* Some dried materials—for instance, the plume grass—could begin to shed while you work with them. Simply spray them with some hairspray to help lock them all into place. The more "dated" the hairspray, the better. Dried materials love the brands of hairspray that were especially popular in the 1980s and '90s, like Aqua Net!

* Creating the base of the design into the curly willow bouquet while it's in your hand in the beginning allows for more control while establishing the shape.

**10.** Thread, or lace, the magnolia stems evenly throughout the curly willow kubari to start your base.

**11.** Add the bleached bell cups in a grouping to the lower right of the bouquet.

**12.** Add the angel breath and limonium evenly throughout.

**13.** Now add the foam flowers, playing with their placement and the surface area they stand in to create depth and interest. Wrap the base of the stems with bind wire to hold them in place.

**14.** Place the base bouquet into the cone. Add a cluster of hanging amaranthus, allowing it to drape gracefully over the front left of the cone. If the stems need to be longer or if the stems need extra support, use floral tape and bind to a plant stake or branch extension.

**15.** Add the feathers in an asymmetrical grouping to the upper left.

**16.** Tape the bunny tail stems into small bundles with floral tape and add throughout the design.

**17.** Add the plume grass and the preserved eucalyptus throughout, keeping both long to add an element of line.

**18.** Finish the composition with a backing of the lunaria and a branch or two of curly willow extended from the top left.

# Dusty Lace

Designing flowers seems like an absolute mystery to so many beginners. They often start with their focal flowers and work their way through their materials until they're left with only greenery. They proceed to tuck that around the collar, call it done, and stand back and tilt their head in wonder about why it didn't come together as they imagined it would. The solution to this debacle is to start with the greenery, or some subordinate material that will act as a framework, thereby holding all of your focal flowers exactly where you put them as the arrangement comes together. Threading the flowers through this framework is called lacing, and it is the jumping-off point from being a garden hobbyist arranger to a real-deal floral designer.

## MATERIALS

- Lacey dusty miller
- 1 large and 3 miniature silver/ gray echeveria
- Silver dollar eucalyptus
- White ranunculus
- Low, rectangular glass vase with wooden tray base
- Gray spray paint
- 18-gauge stub wire
- Floral tape

### Tech Talk

* The silhouette created within the rectangular *form* of this design is a skyscraper effect. This can be a really fun and pleasing design to create and to enjoy.

* The way that the ranunculus is *repeated* and arranged in this vase creates tremendous *rhythm*.

**1.** Place the vase in the tray, fill with water, and begin to design into it. Start with the lacey dusty miller. After stripping away any lower leaves from the stem that might otherwise end up in the water, trim and add to the vase so that the bulk of the stem is resting just above the rim. When you place your next stem, intentionally place it through the lower laterals from the first stem. The lower laterals are the secondary branches that grow from the main stem. Continue to do this, intermingling the stems into each other as you create a dense, but even, framework in the vase.

**2.** Trim the shorter laterals off the main stem of the silver dollar eucalyptus, clearing the bottom of the stems of any leaves. Crisscross these stems into the dusty miller. This adds to the support structure and to the variation in color and texture of the foliage.

**3.** Add the ranunculus stems throughout the dusty miller framework at varied heights and depths. Although the ranunculus are heavy, they are supported and kept upright and in place by the framework.

**4.** To prepare the echeveria for adding into the vase, we need to add a "stem" onto their heads. Trim the echeveria at soil level. Tape the center 3" of two 9" pieces of 18-gauge wire together with floral tape, building a strong but narrow "stem." Add a dab of glue to the end of the wires and pierce straight into the bottom of the echeveria stem. You will feel resistance once you get the wire far enough in, generally 1"–2" is far enough.

**5.** Repeat this step with single pieces of wire for the miniature succulents. Now, these can be placed into the framework. Because there is not a significant color difference between the dusty miller and the succulent, you will get most impact if you add them all in as a group. In this case, we've added them in the center/right, finishing the design.

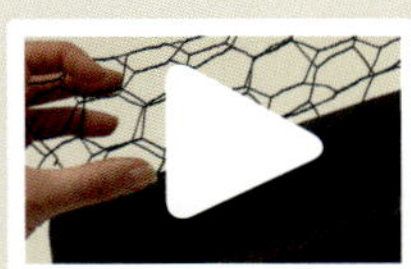
How to change the water for this and other types of arrangements can be found here!

### Tech Tips

* When adding the wire into the stem of the succulent, you'll have the best leverage for solid placement if you turn the succulent over and gently, but firmly, press it in without crushing any of the leaves in the palm of your hand.

* To add a layer of interest to the wooden tray beneath the rectangular glass vase, spray-paint the tray lightly with the gray paint, quickly wiping it with a paper towel. This technique creates a distressed or color-washed effect. You can do this as thick or as thin as you prefer.

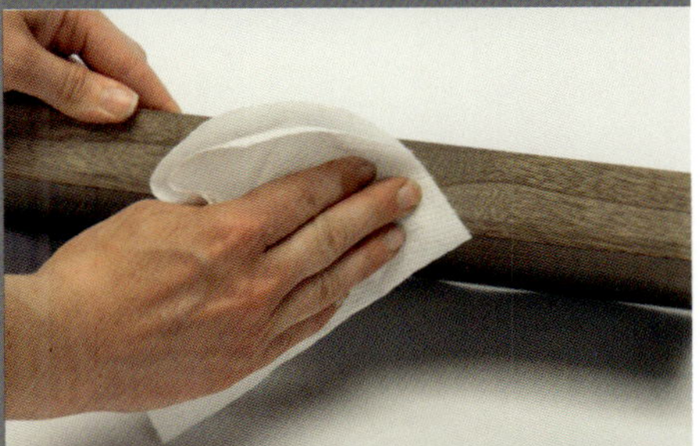

# Mold & Sculpt

This piece is filled with tropical vibes and a touch of the vine. Although you don't typically find flowers from a vineyard climate growing with flowers from a tropical climate, the texture of the grape wood perfectly complements the natural tropical elements. While all of the flowers and foliage in this arrangement are remarkable, the dried cecropia leaves deserve a special highlight. The two-toned nature of these leaves makes them versatile for many color palettes. They are dried, but they aren't as brittle as some dried materials. The arching stems of the massaged anthuriums spring from the cecropia leaves, leading the eye into the phalaenopsis orchid focal area in the center.

## MATERIALS

- African mask alocasia leaves
- Dried cecropia leaves
- Grape wood
- Green dendrobium orchids
- Green obake anthurium
- Mojito green hydrangea
- Palm leaves
- Philodendron leaves
- White cymbidium orchid
- White phalaenopsis orchids
- Large basket
- A fitted liner or combination of containers to fill the basket
- Floral foam
- Bullion wire, any color
- 18-gauge stub wire

### Tech Talk

* If the anthurium petals are tight and turned in toward the spadix, gently pull the colored spathe back and *reflex* it to open the face.

* There is no question that the *focal area* in this design is front and center, made from the dripping white phalaenopsis orchids. The green dendrobium orchids, cymbidium orchids, and anthurium all around the phalaenopsis seem to be almost pointing at them. They are the star of this show!

**1.** Filling this basket completely with wet floral foam would make it very heavy. To work around that, fill the center with an upside-down bucket. Top it with a 12" plastic saucer and wedge a wet Grande block of floral foam into the saucer.

**2.** Hold the grape wood where you think you'll want it to reside within the design, score the outline of it on the foam, then cut out a section of the foam for the grape wood to nestle into.

**3.** Rest the grape wood into the cavity you've created. To make sure it is secured, bend an 18-gauge wire in half and press it into the foam to help anchor it. Repeat this with a second piece of wire.

**4.** Massage the stems of the anthurium by firmly running your thumb along the stem, applying gentle pressure and the warmth of your hand as you go. Repeat this a few times per stem; each time you will see more of an arch develop. Add a grouping of the anthurium into the foam on the right, allowing the new arch that you created to really carry the blooms outward.

**5.** On the left, add a stem of white cymbidium orchids with their petals reflexed.

### Tech Tips

* Manipulating the fronds of the palm leaves into a nautilus shape is commonly done in contemporary tropical-styled design. The fronds are usually pulled into the shape all the way through the tips of palm leaves, but you can also leave the tips naturally upright and add the nautilus effect only to the bottom halves of the palms, as has been done here.

* It's important in a design like this with a lot of drape that the flower foam is about 2" taller than the rim of the basket so that you can add the flowers on an inverted angle as needed.

* Enhance the grape wood with a light airbrushed layer of white paint to bring it more into tone with the whitewashed basket.

**6.** Add a few stems of bright-green hydrangea low and in the center, around where the grape wood meets the foam. Be sure to get a sharply angled cut on the hydrangea and really sink them into the foam. Begin to establish balance with palm leaves placed in the back left, behind the cymbidium orchid.

**7.** Drape the white phalaenopsis orchids from the hydrangea area and allow it to spill over the front of the basket.

**8.** Add the cecropia leaves behind the anthurium to provide a solid backdrop for them. Now place the alocasia leaves under the orchids, dripping down the front of the basket, and the philodendron leaves to their left.

**9.** Add some palm leaves around the anthurium with the bottom ⅔ of their leaves stripped for a lighter touch. A grouping of the green dendrobium orchids on the left balances the spray of green palm and anthurium on the lower right.

**10.** Prepare 3 palm leaves that have partially had their leaves gathered and bound with bullion wire a third of the way down the stem to create interest. Place your gathered palm leaves into the design behind and to the right of the anthurium.

 Scan here to see how easily the nautilus leaf takes shape.

# Slate Shadows

Sometimes the occasion calls for a calming design. One composed of a few well-placed materials. The duo of aspidistra leaves on the left casts a three-dimensional shadow on the trio of luscious roses in the center. Those roses are peacefully backed with a pair of dried fan palms in a coordinating tone, and the eye falls softly down the visual path that the rolled aspidistra leaves create on the right. Not only is this a calm and peaceful design, it is also incredibly long-lasting, which is a bonus for the receiver!

## MATERIALS

◇ Aspidistra leaves

◇ Cream roses

◇ Dried fan palms

◇ Shallow, wok-style bowl

◇ Flower foam

◇ Gray marble chips

### Tech Talk

* The aspidistra leaves on the right and the left, while used in different ways, are *framing* the roses.

* The marble chips on the foam is another example of *basing*.

* The gentle triangle *form* of the container is repeated in the trio of roses.

**1.** Prepare the container with wet floral foam that nestles into the bowl. Bevel out the sharp edges on the top of the foam.

**2.** Massage an aspidistra leaf by rolling the tip in and around your hand, holding it there for a few seconds while the warmth of your hand helps it stay that way. Then place the stem in the foam. Repeat with the second aspidistra leaf, placing its stem just beneath the first. This creates the arrangement's first shadowing example.

**3.** Check your roses over for any bruised or damaged petals that might need to be discarded, then place them in a triangle into the center of the design.

**4.** Add 2 dried fan palm leaves, behind the roses, positioned so that one shadows the other.

**5.** Roll the tip of an aspidistra leaf back onto itself and use a single staple to secure it, as shown in the Sheltered Sanctuary project (page 156). Repeat this with the second leaf. Place both in the foam to the right of the roses, again the second being placed just beneath the first.

**6.** Sprinkle the marble chips on and around the foam, filling the base.

### Tech Tip

Press the marble chips into the foam gently to get them to take hold and keep them from falling out if the piece is moved.

# Tailored Tulips

Contemporary, modern ART! This combination of tailored (slightly modified) foliage, combined with tulips that have been allowed to be open and in their full glory, is perfect for an opening gala at an art gallery. Done in fresh material, it will last a few days. However, this impressive design could easily be replicated with permanent botanicals.

## MATERIALS

- Bear grass
- Equisetum
- Green Spanish moss
- Israeli ruscus
- Monstera leaves
- Umbrella papyrus
- Variegated pittosporum
- White tulips
- Wall planter
- Midnight foam
- UGlu dashes
- Pearl-headed pin
- Straight pins

### Tech Talk

* While each foliage block is different, the tulips used throughout the piece create *unity*.

* Each block carries its own unique *repetition*, which must be considered when placing the materials. Even small details, such as the pins holding the equisetum in place, should be placed with consideration for how it will be viewed and appreciated.

**1.** We'll look at the foliage blocks first, filling every other block on the left, and the alternating blocks on the right. Cut floral foam into blocks to fit each section's length, width, and depth dimensions.

**2.** Fit the wet foam blocks into place.

**3.** Grasp the leaves of the papyrus and gather them together. Clip the leaf tips off, creating a blunt edge, leaving the overall length of the leaves to be 3"–4". Insert them into the first tile of the composition.

**4.** Trim the individual ruscus leaves from the main stem. Curl each leaf into a cylinder shape, securing with half of a UGlu dash. Thread a pin into the center of the curl and pierce it through the overlapped section of the leaf where the UGlu is. Use the pin to hold the ruscus in place in the foam, placing these curls with purpose and repetition.

**5.** As you place these curls, you may find it aesthetically pleasing to nestle the top of the lower row of curls into the bottom of the top row. Roll the ruscus looser or tighter as needed to keep a uniform appearance.

**6.** Gather a bundle of 6–7 pieces of slightly dehydrated bear grass. Roll the tips and create a knot. Then create a second knot, as close to the first as possible. Do this 3, 4, or 5 times per bundle of grass, making a stacked knot, but still also leaving enough tail in the base to add into the foam.

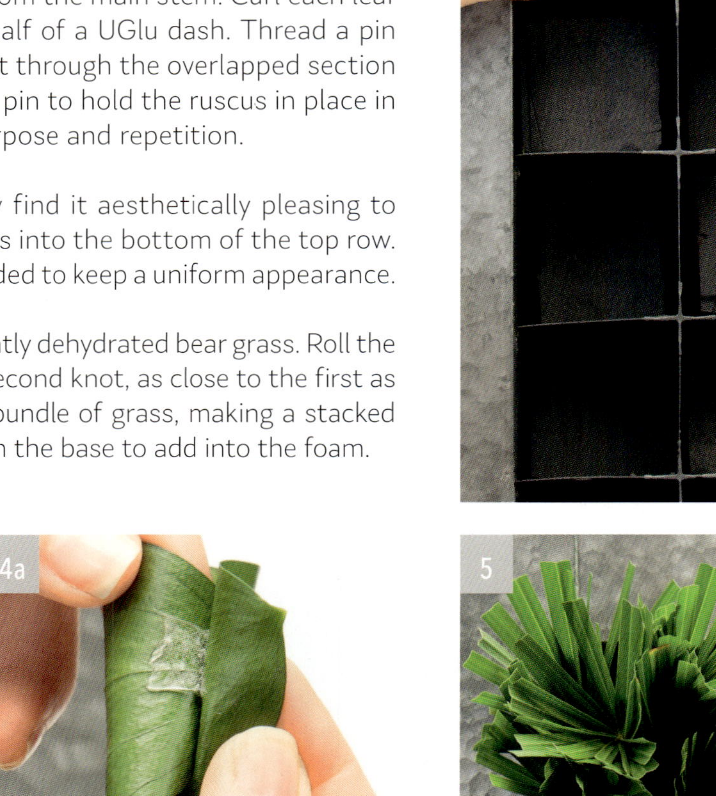

**7.** Place the bear grass bundles evenly into the foam.

**8.** For the fourth foliage tile, apply floral glue to the backs of the pittosporum leaves that have been cut from the main stem, smearing it along the central vein area. Glue them into a circular rosette on the foam surface.

**9.** Make a knife slit in the center of the rosette and insert a tuft of pittosporum from the tip of the stem to serve as the center of this fantasy flower.

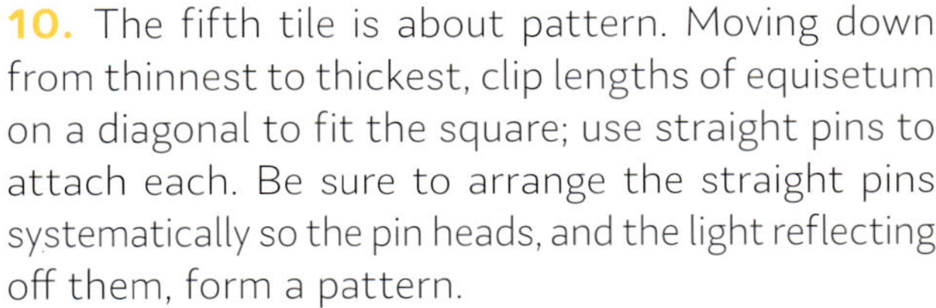

**10.** The fifth tile is about pattern. Moving down from thinnest to thickest, clip lengths of equisetum on a diagonal to fit the square; use straight pins to attach each. Be sure to arrange the straight pins systematically so the pin heads, and the light reflecting off them, form a pattern.

**11.** From the center, begin graduating back down to thinner equisetum lengths.

**12.** For the lower right tile, insert a medium-sized monstera leaf so that it lies against the zinc. Take a second monstera leaf and curl each of its leaf segments inward, attaching the curl with a UGlu dash.

**13.** Add the curled monstera leaf just above the first one.

**14.** For the 6 tiles that will contain tulips, add a thin layer of Spanish moss to cover the surface.

**15.** Gently reflex some of the tulips' petals as shown in the Open Orchids project (page 128).

**16.** Add the tulips into the moss blocks, varying the stem lengths for depth. For detail and variation, include some tulips that are only partially reflexed, with their inner petals still in place, and some that are not reflexed at all.

### Tech Tips

* Tulips are phototropic. They will continue to grow toward a light source even after they have been cut and placed in the design. This needs to be compensated for in terms of timing. If the composition will be used the next day, for example, you'll compensate in terms of stem lengths.

* The tulips will reflex only if they are at room temperature. If they are cold, the petals will snap off, or you may see some split in the center of the petal. The split in the petal can lead to its own intriguing effect and is sometimes encouraged, as it was in this design.

12b

13

14

15

16

# Wrapped Wreath

This styled and chic wreath comes together very easily by tailoring the leaves with a wrapping technique. Utilizing the naturally variegated lines that run down the length of the pandanus leaves, we create the illusion that there are many more materials wrapping this wreath than there are, which increases perceived value. This is also a unique way to design with these leaves, to manipulate them as a wrap instead of in a more traditional sense. The composite rose at the base is an elegant finishing touch. This design could serve as a modern sympathy tribute.

## MATERIALS

- ◇ Pandanus leaves
- ◇ White roses
- ◇ 18" Styrofoam wreath
- ◇ Rustic wire
- ◇ Floral glue
- ◇ 18-gauge wire
- ◇ Pearl-headed pins

### *Tech Talk*

* To make it easier to add the extra petals around the center rose of the composite, do not *reflex* the outer leaves, which will ensure that you have a ledge on the back on which to glue.

* By first curling the pandanus by hand, loosening up the fibers so that it can be pinned easier, the *massaging* technique is being used.

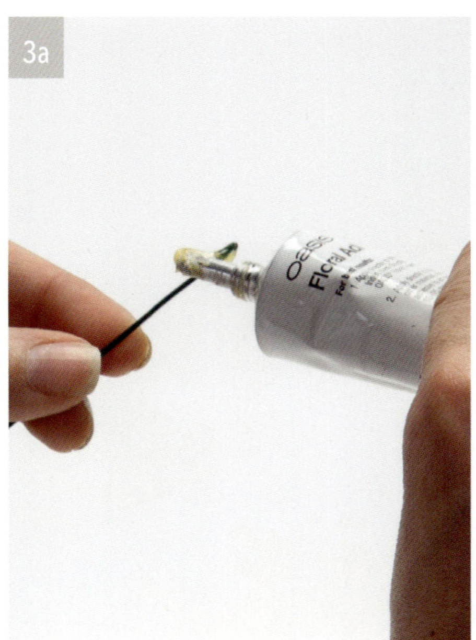

**1.** Create a hook for the back of the wreath with an 8" length of rustic wire by looping it and twisting it several times. Then, pull the natural fibers up from the ends and insert the internal wire into the foam. Add floral glue around the area where the ends went in to the Styrofoam to doubly secure it. Let dry.

**2.** To prepare the composite rose, choose your perfect center rose. Pluck off any outer petals that are less than desirable. Pull off the green sepals just below the rose head and cut the rose so that there is a 2" stem left.

**3.** Run a bead of floral glue down the first inch of a 6" piece of 18-gauge wire. Pierce the wire into the stem of the rose. Gently set the rose on its head.

### Tech Tips

* The pandanus leaves turn brown if put in the cooler. If left at room temperature, they will last for a very long time. Even out of water and manipulated as these leaves are, they last in this application for several days.

* The number of leaves you will need to complete this will depend on the leaves' lengths and widths; this wreath used approximately 20 leaves.

**4.** From a second rose, rock the head off the stem. This will give you the extra petals you'll add to the first rose. Sort through the extra petals to choose only the largest and most beautiful.

**5.** Dry-place each petal before you glue it onto the first rose to understand where the contact points will be. Then, add the floral glue to those spots on the petals. Let the glue get tacky for 10–15 seconds and place around the base of the first rose.

**6.** Begin layering the additional petals around the base of the rose, lightly pressing each on to adhere. Keep in mind the natural structure of a rose's petals, staggering them. For this project, add 12–15 petals.

**7.** To make the wreath, starting with a single pandanus leaf, clip off and discard the rigid lower portion of the leaf. Bend and roll the remaining lower portion of the leaf to make it more flexible.

**8.** Beginning with the cut end, wrap the leaf around the form, bending and creasing as you go. Pin down the tip with a greening pin, or a small U-shaped piece of 20-gauge wire to secure it.

**9.** Repeat this and continue wrapping the wreath until you have gone all the way around.

**10.** Taking only the flexible top 6"–8" of a pandanus leaf, curl the tip around, insert a long pin through the curl from the inside, and pin it to the left side of the wreath, similar to how the ruscus was added in the Tailored Tulips project (page 204).

**11.** Repeat this with the next section of leaf that is left from the tip you just removed. Curl it and pin it into place. Continue cutting segments, moving down the leaf, to create additional leaf curls. You can create 2 to 3 curls from the average leaf before reaching the lower portion that's too stiff to curl. Group the curls closely, to create an effect similar to ribbon loops.

**12.** As you add the curls, you'll position the middle/wider portions' leaf curls below and under the weaker veined curls from the tips to provide support.

**13.** Add the rose by piercing its wire through the leaf wrap to anchor securely into the Styrofoam.

# Dripping Decadence

There are many ways to create a floral chandelier. This one is built on an easily reusable base, with loads of nooks and crannies to add floral material into. Although it's strong enough to have floral foam added to it for an even more lush chandelier, we have created this project using the foam-free method. The highlight of this installation is definitely the lovely mini calla lilies. Even the bright-green hue of the calla lily stems becomes part of the effect. By massaging the stems, we are able to get an even greater arch from the base, creating a shower effect.

## MATERIALS

- Bear grass
- Seeded eucalyptus
- Silver dollar eucalyptus
- Smilax vine
- White mini calla lilies
- 3 assorted-sized narrow grapevine wreaths: 18", 24", and 30"
- 8" and 11" cable ties
- 20-gauge spool wire
- Wood picks
- Pearl-headed pins

### Tech Talk

* Aside from the greenery in this piece, by adding only calla lilies we have made the choice for the composition to be *monobotanical*, meaning that it includes only one type of flower.

* There is an interesting *variation* in this choice of materials due to the silky smilax, the smooth eucalyptus, the airy grass, and the sleek calla lilies.

**1.** To create the grapevine base, set the wreaths on the table in ascending size like a bull's-eye. Create the anchor points by attaching an 8" cable tie to a "corner" of the largest wreath. It does not matter where you start, but this will become the basis for spacing your other points. Pull this cable tie snuggly but leave a bit of room for another tie to slide between it and the wreath. Then, add a cable tie to the other two wreaths at this same point. Repeat this process across the wreath on the opposite corner, then the two side corners. When finished, you will have 12 cable ties attached to the wreaths.

**2.** Now, loop an 11" cable tie through the cable tie on the smallest wreath and the cable tie in the middle wreath. Pull until you get 2–3 clicks in, so there is quite a bit of slack in the tie. Repeat this process between the middle wreath and the largest wreath. Then, repeat it all the way around. When finished, you will have added 8 11" cable ties, and all of your wreaths will be officially joined. Trim all cable tie tails away.

**3.** To create the hanging mechanism, slide the spool wire through one of the outer ring cable ties, pull a tail through that is about 2" long, and twist up the wire to create a tight connection. Without cutting the wire, stretch it across to the opposite corner, estimate the length you will need to repeat the twist process, and the length you will need for actually hanging the structure (always better to cut a little extra), then cut that amount of wire from the spool. Feed the loose end of the wire through the outermost cable tie on the opposite side, fold up, and twist. Repeat this process in the cross section. Using needlenose pliers to work the wire will make the job easier. Pull both wires up to where they meet, create a loop about 2" in diameter, and twist.

**4.** Hang the structure for working.

**5.** Begin by loosely wrapping the smilax vine over the top, down, and through the wreaths.

**6.** To attach the silver dollar eucalyptus, firmly insert the stems into the grapevine wreaths. These stems should hold in place with tension by the grapevine. Add glue if they feel loose.

**7.** Add the seeded eucalyptus by working the stems firmly into the grapevine.

**8.** Once you are happy with your amount of greenery, begin to massage the stems of the calla lilies by gently applying pressure as you run your grip from the base of the stem toward the flower. Massaging accentuates the arch and gives a graceful look. The more you massage, the more the arch emerges.

**9.** Insert a pearl-headed pin in the base of each stem. The pin gives the soft stem reinforcement and leverage to allow you to insert it into the grapevine.

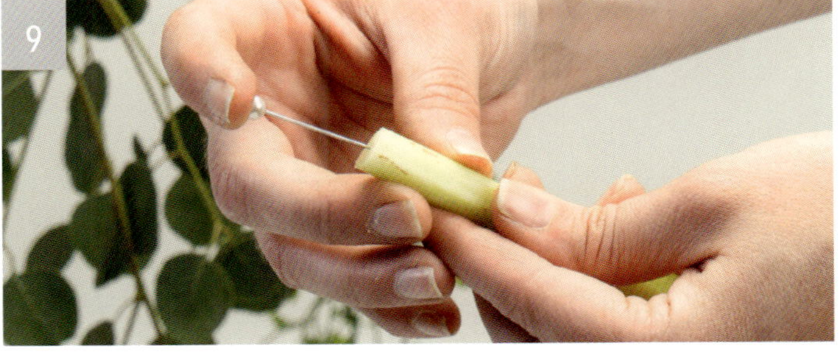

**10.** Place the calla lilies into the wreaths. For these, you may need to gently pull the grapevine strands apart a bit more with your fingers to allow space for the stems.

**11.** Gather the bear grass into small groups and attach the stem end to wood picks. Massage the grass by wrapping it firmly around your hand, allowing the movement and your hand's warmth to relax the grass fibers. Add the grass bundles all around, finishing the chandelier.

### Tech Tips

* Dehydrate both the calla lilies and the grass before massaging them by taking them out of water for a few hours before you plan to design. Dehydration will help the stems become more pliable.

* To help keep the wreaths as even as possible, pull the 11" cable ties the same number of "clicks," 2 to 3, at each connection. This way, your base will be as level as possible.

* Focus on adding the greens around the cable tie areas first to be sure this mechanic is covered, then work your way evenly around.

## Index of Techniques



## Materials Sponsor Acknowledgment

Many thanks to Accent Décor for their endless support of education and artistry in the floral industry. Their brilliant products can be seen in the photographs throughout the pages of this book.

ACCENT DECOR

*When it seems that the world can be full of so much of the same, Accent Décor finds great value in the power of new inspiration. From our fresh and innovative products, to our commitment to quality and excellence, we seek to offer design-focused vessels that provide continual inspiration for the floral, event, and home décor industries.*

## Accent Décor Container Index

# Fresh Flower Botanical Name Index

| Common Name | Botanical Name |
| --- | --- |
| African mask alocasia | *Alocasia amazonica* |
| Agapanthus | *Agapanthus africanus* |
| Alocasia | *Alocasia longiloba* |
| Anthurium | *Anthurium andraeanum* |
| Aralia | *Fatsia Japonica* |
| Aspidistra | *Aspidistra elatior* |
| Astilbe | *Astilbe japonica* |
| Bear grass | *Xerophyllum tenax* |
| Bells of Ireland | *Moluccella laevis* |
| Blueberry | *Vaccinium corymbosum* |
| Buplurem | *Bupleurum chinense* |
| Butterfly ranunculus | *Ranunculus eschscholtzii* |
| Button pom chrysanthemum | *Chrysanthemum x morifolium* |
| Calla lily | *Zantedeschia rehmannii* |
| Carnation | *Dianthus caryophyllus* |
| Craspedia | *Craspedia globose* |
| Cremone | *Chrysanthemum × morifolium* |
| Curly bamboo | *Dracaena sanderiana* |
| Curly willow | *Salix matsudana* |
| Cymbidium orchid | *Cymbidium 'Jung Frau'* |
| Daisy chrysanthemum | *Leucanthemum* |
| Dendrobium orchid | *Dendrobium 'Burana Jade'* |
| Dubai ornithogalum | *Ornithogalum dubium* |
| Elaeagnus | *Elaeagnus pungens* |
| Equisetum | *Equisetum hyemale* |
| Eucalyptus parvifolia | *Eucalyptus parvula* |
| Fan palm | *Livistona chinensis* |
| Feverfew | *Tanacetum parthenium* |
| Flat fern | *Nephrolepis exaltata* |
| Freesia | *Freesiaecorymbosa* |
| Galax | *Galax urceolata* |
| Garden rose | *Rosa 'White O'Hara'* |
| Gerbera daisy | *Gerbera jamesonii* |
| Ginger | *Alpina purpurata* |
| Gladiola | *Gladiolus hortulanus* |
| Green Trick dianthus | *Dianthus barbatus 'Green Trick'* |
| Hanging green amaranthus | *Amaranthus caudatus <Viridis'* |
| Heliconia | *Heliconia orthotricha* |
| Hyacinth | *Hyacinthus orientalis* |
| Hydrangea | *Hydrangea macrophylla* |
| Hypericum berry | *Hypericum androsaemum* |
| Iris | *Iris hollandica* |
| Israeli ruscus | *Ruscus hypoglossum* |
| Italian ruscus | *Ruscus aculeatus* |
| Ivy | *Hedra helix* |
| Lacey dusty miller | *Jacobaea maritima* |
| Larkspur | *Consolida ajacis* |
| Lavender | *Lavandula angustifolia* |
| Hybrid lily | *Lilium* |
| Leatherleaf fern | *Rumohra adiantiformis* |
| Lily grass | *Liriope muscari* |
| Lisianthus | *Eustoma grandiflorum* |
| Maidenhair fern | *Adiantum aleuticum* |
| Matsumoto aster | *Callistephus chinensis* |
| Ming fern | *Asparagus macowanii* |
| Miniature carnation | *Dianthus caryophyllus* |
| Mint | *Mentha piperita* |
| Miscanthus | *Mentha piperita* |
| Monstera | *Monstera deliciosa* |
| Nerine lily | *Nerine sarniensis* |
| Nigella pod | *Nigella damascene* |
| Ninebark | *Physocarpus opulifolius* |
| Oncidium orchid | *Oncidium altissimum* |
| Ornamental pineapple | *Ananas comosus* |
| Palm leaves | *Dypsis lutescens* |
| Pandanus | *Pandanus baptisii* |
| Pennycress | *Thlaspi arvense* |
| Peony | *Paeonia 'Riches & Fame'* |
| Persian cress | *Lepidium sativum* |
| Phalaenopsis orchid | *Phalaenopsis amabilis* |
| Philodendron | *Philodendron bipinnatifidum* |
| Phlox | *Phlox paniculate* |
| Pincushion protea | *Leucospermum cordifolium* |
| Plumosa | *Asparagus setaceus* |
| Psittacorum | *Heliconia psittacorum* |
| Pussy willow | *Salix caprea* |
| Queen Anne's lace | *Daucus carota* |
| Ranunculus | *Ranunculus asiaticus* |
| Rose, assorted varieties | *Rosa* |
| Rosemary | *Salvia Rosmarinus* |
| Scoop scabiosa | *Scabiosa columbaria* |
| Seeded eucalyptus | *Eucalyptus polyanthemos* |
| Silver dollar eucalyptus | *Eucalyptus cinereal* |
| Smilax | *Smilax smallii* |
| Snapdragon | *Antirrhinum majus* |
| Spray rose, assorted varieties | *Rosa* |
| Sprengeri | *Asparagus aethiopicus* |
| Spurflower | *Plectranthus hadiensis* |
| Sunflower | *Helianthus annuus* |
| Ti leaves | *Cordyline fruticose* |
| Tillandsia Xerographica | *Tillandsia xerographica* |
| Trachelium | *Trachelium caeruleum* |
| Tree fern | *Asparagus virgatus* |
| Tulip | *Tulipa gesneriana* |
| Uluhe fern | *Dicranopteris linearis* |
| Umbrella papyrus | *Cyperus alternifolius* |
| Variegated flax | *Dianella tasmanica* |
| Variegated lily grass | *Liriope muscari 'Variegata'* |
| Variegated pittosporum | *Pittosporum tobira 'Variegata'* |
| Veronica | *Veronica officinalis* |
| Waxflower | *Chamelaucium uncinatum* |

## Plant Botanical Name Index

| Common Name | Botanical Name |
| --- | --- |
| Blue star fern | *Phlebodium aureum* |
| Cactus | Various members of the Cactacae family |
| Stonecrop | *Crassula* |
| Croton | *Codiaeum variegatum* |
| Echeveria, assorted varieties | *Echeveria* |
| Grafted moon cactus | *Gymnocalycium mihanovichii* |
| Inchplant | *Tradescantia zebrina* |
| Kalanchoe | *Kalanchoe blossfeldiana* |
| Maidenhair fern | *Adiantum capillus-veneris* |
| Nandina | *Nandina domestica* |
| Rex begonia | *Begonia rex-cultorum* |
| String of pearls | *Senecio rowleyanus* |
| Oat grass | *Avena sativa* |
| ZZ plant | *Zamioculcas zamiifolia* |